The Mountainside Breeding Herd of Jersey Cattle
of Mahwah, New Jersey

by Theodore A. Havemeyer

with an introduction by Jackson Chambers

This work contains material that was originally published in 1883.

This publication is within the Public Domain.

This edition is reprinted for educational purposes and in accordance with all applicable Federal Laws.

Introduction Copyright 2017 by Jackson Chambers

Self Reliance Books

Get more historic titles on animal and stock breeding, gardening and old fashioned skills by visiting us at:

http://selfreliancebooks.blogspot.com/

Introduction

I am pleased to present another title in the "Cattle" series.

The work is in the Public Domain and is re-printed here in accordance with Federal Laws.

As with all reprinted books of this age that are intended to perfectly reproduce the original edition, considerable pains and effort had to be undertaken to correct fading and sometimes outright damage to existing proofs of this title. At times, this task is quite monumental, requiring an almost total "rebuilding" of some pages from digital proofs of multiple copies. Despite this, imperfections still sometimes exist in the final proof and may detract from the visual appearance of the text.

I hope you enjoy reading this book as much as I enjoyed making it available to readers again.

Jackson Chambers

STATEMENT.

The herd of Jerseys herein catalogued has been collected by purchase since the summer of 1880. The animals were bought, with few exceptions, either upon the Island of Jersey or at public auction in this country. The exceptions are a few cows from the fine herd of the proprietor's father, Mr. F. C. Havemeyer, of Throgg's Neck, Westchester County, and a few purchased in dam.

In establishing the herd, the object was to obtain one hundred cows of the highest possible excellence, judged by their milk and butter yield, their fitness for breeders (constitution), and their typical excellence as Jerseys, and neither time, pains, nor expense, which could reasonably contribute to this end, have been spared.

In order that the progeny of these cows shall be, if possible, better than their dams, the services of the best bulls which can either be purchased or hired are secured.

Mountainside Farm is conducted strictly as a business enterprise, in which the proprietor takes lively interest and pleasure. The only products from which income is derived or anticipated are the products of the dairy (including pork and poultry, largely fed upon milk) and the increase of the herd.

This entire increase is offered annually for sale at auction.

No sales whatever are made in any other way, except to the butcher.

[OVER]

The need of a young bull may occasionally, or possibly uniformly, lead to the temporary withholding of one or more for use in the herd.

*Breeders, wishing to purchase, have therefore the assurance—*1st, *that no animal deemed unworthy of the herd will be offered;* and, 2d, *that none, however excellent, bred at Mountainside, will be permanently withheld from the annual sales.*

MOUNTAINSIDE FARM *is in the Valley of the Ramapo, about* 30 *miles from New York, on the Erie Railway, two miles from* MAHWAH, N. J., *and three miles from Suffern, N. Y. The Post-office and Telegraph-station are at Mahwah.*

The Farm is always open to the inspection of visitors, and breeders of Channel Island cattle are especially welcome.

ITALIC LETTERS *are used uniformly in this Catalogue for animals bred in the Island of Jersey (or in England).*

NUMBERS *not in parenthesis refer to the Herd Register of the Am. Jersey Cattle Club.*

NUMBERS *in parenthesis, and accompanied by letters, refer to the Jersey Herdbook, unless marked Eng. H. B. The accompanying letters are:* F., *for foundation stock;* P., *for pedigree stock;* H. C., *for highly commended; and* C., *for commended—the latter being only occasionally used, because all animals admitted to the Jersey Herdbook must be at least "* commended."

INDEX.

Name	Page	Name	Page	Name	Page
Alcmena, 6193	76	Compo, 2d, 14885	95	Kate le Brocq, 14615	98
Alcmena, 3d, 14614	92	*Coomassie*, 2d, 11969	30	*Lady Arthur*, 11966	28
Amulet of Home Farm, 12096	70	*Cream of Jersey*, 18181	111	Lady Hammond, 7286	74
Amy la Grise, 11952	18	*Crescent of Mapleshade*, 10927	38	Lady Morgan, 11671	62
Anna of Mountainside, 15544	53	Dairy Maid of Bloomfield, 8352	59	*Lady Pert*, 14613	96
Annie Golddust, 6849	78	Deerfoot Girl, 15329	114	Lady Sygny, 6938	55
Art Souvenir	86	Empressa, 4th, 12278	42	*La Financiere*, 11970	31
Arthur's Mistletoe, 11968	29	Fairy Gold Dust, 14612	97	*La Gallichaume*, 15542	51
Bay Queen, 11145	73	*Falladot*, 11967	32	*Landseer's Dove*, 19648	85
Beacon Lass, 14884	99	Fancy Fan, 12657	100	*Lille Bonne*, 8108	79
Beauty of Ninon, 9691	57	*Fauvette*, 15021	44	Lille Bonne, 2d, 12253	94
Belle Dame, 11951	36	Farmer's Glory, 5196	8	Lily of Cedar Grove, 13912	104
Bergerelia, 15546	50	Favorite's May, 6662	67	Lindora, 5361	37
Bijou le Cerf, 11964	21	*Fleur de Leury*, 19651	80	Lola Hammond, 12252	93
Brownie, 1166	71	Flora D., 11192	64	*Mahwah Rose*, 21708	90
Browse, 14611	91	Gambette, 1 541	45	Maid of the Valley, 16556	102
Brunette Hammond, 7284	77	Gay Grisette, 9991	54	Malita, 5169	68
Brunette le Briton, 11955	23	Glorinessa, 18441	27	Malvolio, 5568	11
Carlo, 5559	10	Golden Touch, 18221	113	Mantissa, 10457	61
Carlotta, 18533	103	Hilpa, 5879	56	*Mary Jane*, 18185	112
Casta Cazique, 16963	108	*Island Flower*, 15265	106	*Maud*, 11961	35
Cocotte, 11958	24	*Jersey Rose*, 18216	110	*Mignonne*, 11959	25
Colt's La Biche, 6399	105	*Jessy Belle*, 19647	81	*Miss Alexandre*	83
Compo, 11844	109	Jessy of Ipswich, 2883	58	Miss le Feuvre	89

Continued.

	PAGE		PAGE		PAGE
Miss Huelin	84	Phyllis, 10657	66	*Stolen Kisses*, 16864	101
Mouse, 11953	16	Polly of Deerfoot, 15328	115	*St. Ouen*, 11963	22
Nameless Girl, 11623	63	Pride of Mountainside, 7118	12	*St. Martainaise*, 15543	46
Nelly of Mountainside, 15545	47	*Primrose*, 11956	33	*Sultane* 4th, 11960	34
Nellie Rival, 11147	72	*Princess*, 11957	26	Thorndale Belle, 2d, 6421	60
Northern Queen, 11962	19	*Redbreast*, 15022	49	Token, 10262	65
Nymph of St. Lambert, 12968	107	*Romulus' Lively*, 15024	48	Troth Plight, 10258	40
Ochra, 4845	75	*Rosa*, 11954	17	*Utalpia*, 19650	87
Pansy, 11965	20	*Royalist's Gypsy*, 15023	52	*Victor's Lucie*, 19649	82
Paraphrase, 10254	39	Satin, 10329	69	Witch of the Maples, 10093	41
Parasol, 12658	43	Silver Sheen, 9047	13	*Young Mouse*, 21493	88

Index to Famous Animals represented by Progeny in this Herd, on pages 116 and 117.

BULLS.

FARMER'S GLORY, with Portrait, pages 8 and 9.

CARLO, with Portrait, page 10.

MALVOLIO, *with Portrait, page 11.*

PRIDE OF MOUNTAINSIDE, *page 12.*

SILVER SHEEN, *page 13.*

A Portrait of FARMER'S PRIDE (son of FARMER'S GLORY), sold at auction, May, 1883, is given in connection with that of his dam, facing page 25.

I. *Farmer's Glory (F. 274 H. C.), 5196.* Solid gray, shading to buffalo fawn, with black tongue and switch. Long-bodied and straight; very deep and broad, with wide hips and long, broad rump. Head long, broad between the eyes, which are full and large, face deeply dished; horns strong, incurving, waxy, and rich; neck masculine; crops full; ribs well sprung; loins wide; flanks deep; tail fine and sweeping the ground; twist moderately low, and thighs thick; legs straight and fine below knees and hocks; hide mellow, loose, and unctuous; coat soft, and skin color rich yellow; escutcheon broad, large in area, not rising above the twist. Disposition timid; carriage noble and up-headed.

History—Dropped March, 1878. Bred by Francis le Brocq, St. Peter's Parish, Jersey. Qualified for registration in the Island Herdbook, April, 1879. After winning many prizes, enumerated below, was sold to Thomas H. Dudley, of Camden, N. J., and imported to Philadelphia in the steamship Lord Gough, August 12, 1880, and led his fine herd for about one year.

Prizes Won: 1st over all Jersey, 1st St. Peter's, and 1st at "The Royal," Kilburn, England, in 1879; 1st over all Jersey, 1st St. Peter's, 1st at Pennsylvania State Show, Philadelphia, and 1st in his class, and also 1st as leading the Prize Herd at the New Jersey State Show at Waverly, in 1880, since which time he has not been exhibited.

He was bought by Mr. Havemeyer at the sale of Mr. Dudley's herd, in connection with the first instalment of the Cooper-Maddux "Oxford Park" Herd, December 1st, 1881, for a price then higher than any Jersey bull ever sold for. At the same sale one of his sons, *Forget-me-not*, sold for $2,125; four of his sons averaged $882.50 each, and fifteen heifers averaged $477.34 each, and at the sale of Mr. Wing's Mapleshade Herd, May, 1881, three heifers, his daughters, sold at an average of $827 each.

[PEDIGREE ON FOLLOWING PAGE.]

Pedigree

FOLLOWING THE LINE OF SIRES.	COLOR.	BREEDER.	DAM.	COLOR.	SIRE'S BREEDER.
Farmer's Glory (F. 274 H. C.), 5196.	Gray	F. le Brocq	*Bonheur* (F. 1651 H.C.)	Lt. br'n	F. le Brocq.
Sire, *Grey King* (P. 169 H. C)	Silver gray.	Wm. Alexandre	*Lily Grey* (F. 770)	Silver gr.	Wm. Alexandre.
2d Sire, *Duke* (P. 76 H. C.)	Gray	Clement Lesbirel.	*Superb* (F. 353 H.C.)	Br. & w.	C. Lesbirel.
3d Sire, *Merry Boy* (P. 61 H. C.)	Gray	John Arthur	*Eva* (F. 628)	Br. & w.	J. Arthur.
4th Sire, *Stockwell*, 2d (P. 24 H. C.)	Yel. & w.	J. C. Godfray	*Soucique* (F. 68)	Y. & w.	Ph. Godfray.
5th Sire, *Noble* (F. 104 H. C.)	Lt. brown	C. Pallot	——	——	——
6th Sire, *Sultan* (F. 58 H. C.)	Lt. brown	C. Pallot	*Flower* (F. 53 H.C.)	Br. & w.	C. Lesbirel.

Pedigrees following the line of sires are rarely found except among Jersey-bred animals. Island breeders send cows of the *Foundation Stock* to prize-winning bulls; hence these pedigrees. In this case, these are all highly commended, all prize-winners, and the dams are or were among the most famous on the Island.

GREY KING won first over all Jersey, and third at "The Royal," Bristol, England, in 1878.

DUKE, 76—called in this country "Sweepstakes Duke," 1905—won first over all Jersey, and first Parochial (Trinity Parish) and first Herdbook prizes in 1875.

STOCKWELL, 2d, won the Bronze Medal at the Channel Island Exhibition in 1871, the second Herdbook prize, and the third prize over all Jersey in 1871, and the third over all Jersey in 1872.

NOBLE won first Parochial (St. Saviour's) and second prize over all Jersey in 1869.

BONHEUR won, as a two-year-old heifer, first at St. Peter's Parochial, and first as a three-year-old at the same show, second (Young Garenne taking first) over all Jersey in 1881, and first and sweepstakes at the show of the three Western parishes the same year.

It is no wonder that the get of *Farmer's Glory* prove so good. Mr. C. F. Dorey, Secretary of the Jersey Herdbook, writes in 1881 that never to his knowledge had so many prizes been awarded to the offspring of one bull.

The sons of *Farmer's Glory* head several of the best herds in this country and England, and his daughters exhibit almost uniformly beauty of form, great style, exquisite "touch," beautifully shaped udders, broad escutcheons, richness and high average quantity of yield.

II. *Carlo (P. 180 H. C.), 5559.* Solid orange fawn, with silvery gray shadings, dark head and legs; full, black points, tongue and switch. A large bull, low-set and long-bodied, exceedingly level and deep, broad-loined, wide-hipped, long-rumped, square, and unusually deep in the flank; head light and fine; face very dishing; eyes wide-set, full, and mild; horns incurving, set forward, exceedingly waxy and yellow; neck masculine, but light; crops full; brisket low; chest broad; thick through the heart; nether line nearly level; tail fine and long; legs fine-boned and straight; twist full and deep; hide of medium thickness, loose, mellow, and unctuous; coat short and silky; skin surface as full of rich, yellow yolk as a Guernsey's, especially abundant in the ears; rudimentary teats, long, uniform in size and squarely placed. In disposition he is gentle and trustful, and though prompt and active, lacks style.

History—Dropped February 11, 1877. Bred by Thomas Falla, Jr., St. John's Parish, Jersey. Qualified August 4, 1878. Bought of his breeder, by E. Burnett for T. A. Havemeyer, in August, 1880, and shipped in September of the same year to New York in steamship Marengo, from Southampton.

Prizes Won : As a yearling, 1st St. John's Parochial, in 1878; as a 2-year-old, 2d over all Jersey in 1879, scaling equal in points to the first; as a 3-year-old, in 1880, 1st over all Jersey, scaling 92 points.

Pedigree

FOLLOWING THE DAM'S LINE.	COLOR.	BREEDER.	SIRE.	COLOR.	SIRE'S BREEDER.
Carlo (P. 180 C.)..............	Orange f'n.	Thos. Falla, Jr.	*Hero* (P. 126 C.)......	Gray.......	Thos. Falla, Jr.
Dam, *Pretty Maid* (F. 1493 C.)..	Brown.. .	Thos. Falla, Jr.	*Yankee* (P. 27 H. C.).	Light br'n..	E. Gibaut.
2d Dam, *Tinker*..............	Dark fawn.	Thomas Falla ..	*Black* (F. 593 H. C.).	Black......	John le Brocq.
3d Dam, *White Rose* (F. 593 C.)	Gray & w.	Thomas Falla ..	————	————	————

HERO (P. 126), *Carlo's* sire, is by *Dick*, a bull of Mr. Falla's breeding, 171 of the *Foundation Stock*, and highly commended. He was by *Yankee* (P. 27 H. C.), the sire of *Carlo's* dam; while *Cowslip* (P. 24), *Hero's* dam, also bred by Mr. Falla, and out of a famous old cow of his father's, *Brown Fanny*, is by John Arthur's *Orange Peel* (F. 129 H. C.), a son of Falla's *White Rose* (F. 593). *Carlo's* short pedigree therefore exhibits two crosses of *Yankee* and two of *White Rose*. He was selected on the Island on account of the reputation of his dam and her line as butter cows, and his heifers show the remarkable certainty with which he imparts butter characteristics.

NERO (P. 248 H. C.), a son of *Carlo's*, won first St. John's Parochial prize in 1880 and 1881, and first over all Jersey in 1882.

A note from *Carlo's* breeder, June 12, 1882, states that in 1880 he was judged by the same men (committee) who judged the two-year-old class, in which the famous *Farmer's Glory* won first, and that he beat the latter by two points, being awarded 92 to *Farmer's Glory's* 90. which number (92) was the highest on record to that date.

III. **Malvolio, 5568.** Fawn and white, the white in spots with flecks of black; white tongue, black switch; back straight, level and long; loin and hips of medium width; rump very long and slightly depressed; crops full; barrel hooped, and close-ribbed back; neck light with little dewlap; head well shaped; horns set wide, strong but rich and waxy; hide mellow and elastic; escutcheon medium; rudimentary teats long and squarely placed; disposition kindly and curious; carriage fine.

History—Dropped October 28, 1880. Bred by Alfred B. Darling. Bought by T. A. Havemeyer at the Kellogg Combination Sale, May, 1881.

Pedigree

	COLOR.	BREEDER.	SIRE.	COLOR.	BREEDER.
Malvolio, 5568	F'n & w.	A. B. Darling.	Duke of Scituate, 3623	Bl'k, gr. & w.	Chas. O. Elms.
Dam, *Grace Darlington*, 5574	Lemon.	Ph. Renouf	*Clement*, on Island of Jersey	———	———
2d Dam, *Violet of Darlington*, 5573	———	Ph. Renouf	*Le Galais, prize bull of* 1867	———	A. le Galais.

DUKE OF SCITUATE is the son of Jersey Belle of Scituate, one of the most famous of Jersey cows (now dead); tested for one year, she yielded 705 pounds of butter.

VIOLET OF DARLINGTON was never thoroughly tested, but gave in her flush 18 quarts of exceedingly rich milk. When giving 15 quarts a day, 12 quarts yielded 2 lbs. 4 oz. of butter. The dam of Malvolio is granddam to the now famous BOMBA.

EAR
| 21-2 |
MARK

Pride of Mountainside, 7118.

Dark gray on back, black on sides, with white spots, white tongue and switch. An exceedingly well-formed bull, combining in a remarkable degree the excellences of both his sire and dam; straight backed, level, and wonderfully deep in body; level below, to the flanks, and broad in loin; hips broad and long to rump. His hide is thin, loose, mellow, and of almost unsurpassed richness of color, his coat soft and fine, and his escutcheon unusually good.

History—Dropped March 6, 1882. Bred by T. A. Havemeyer, Mountainside. Has never left the Farm.

Pedigree

	COLOR.	BREEDER.	SIRE.	COLOR.	SIRE'S BREEDER.
Pride of Mountainside, 7118.	Gr. & w.	T. A. Havemeyer..	Duke of Darlington, 2460.	D'k gray.	A. B. Darling.

Dam, *Belle Dame* (F. 1934 H. C.), 11951, Francis le Brocq.

DUKE OF DARLINGTON is the best son of EUROTAS, through whom he gains the blood of *Rioter, 2d,* famous in its combination with that of Alphea.

BELLE DAME is a highly commended cow of the *Foundation Stock* of the Island. She is No. 21 of the Mountainside Herd, and a cow of remarkable beauty and of the highest quality.

EAR MARK |60—1|

Silver Sheen, 9047.
Solid dark silver gray, black on sides and forequarters, black tongue and switch. A grand bull, very level, wide, and deep, with well-sprung ribs, thick through the heart, square behind, low and close twist; hide mellow and unctuous, coat silky, and skin color very rich.

History—Dropped May 4, 1882. Bred by T. A. Havemeyer. Has never left the farm.

Pedigree

	COLOR.	BREEDER.	SIRE.	COLOR.	BREEDER.
Silver Sheen, 9047..	Solid silver gray.	T. A. Havemeyer.	Bl'k Prince of Hanover...	Dark gray.	A. B. Darling.
Satin, 10329.......	Solid chocolate..	A. B. Darling ...	Duke of Darlington, 2460.	Gray......	A. B. Darling.
Dam, Oriole, 2563..	Dark fawn & w.	R. M. Hoe......	*Dolphin*, 2d, 468.........	Gray......	F. M. Wilson,
2d Dam, Leda, 799.	Solid dark fawn.	R. M. Hoe......	Jupiter, 93..............	Brown & w.	R. M. Hoe.
3d Dam, Europa, 176	Brown fawn....	R. M. Hoe......	Jupiter, 93..............	Brown & w.	R. M. Hoe.
4th Dam, Alphea...	Brown fawn....	R. M. Hoe......	*Saturn*, 94..............	Reddish f'n	R. M. Hoe, impr.
5th Dam, *Rhea*, 166	Dark brown....	R. M. Hoe, impr.			

BLACK PRINCE OF HANOVER has identically the equivalent breeding to EUROTAS, being by *Rioter*, 2d, and out of Leda, a daughter of Europa, by her full brother Jupiter.

DUKE OF DARLINGTON, 2460, is the most famous son of Eurotas, 2454, perhaps the most famous living Jersey. He is also the sire of Bomba, 10330.

RIOTER, 2d, 467, the sire of Black Prince of Hanover and of Eurotas, and *Dolphin*, 2d, 468, the sire of Oriole, 2563—two English bred bulls of the Dauncey blood—are the only out-crosses (and valuable ones they are) in this otherwise "pure Alphea" pedigree.

SATIN is No. 60 of the Mountainside Herd. A cow of remarkable excellence.

ALPHEA, the mother of this family, by G. W. Harris's test, gave a pound of butter to every six quarts of milk, and kept it up all summer, her yield varying from 24 to 18 quarts a day.

COWS OF THE HERD.

(Named alphabetically in the Index, page 5.)

Photographic Portraits, by Schreiber, of the following Cows are inserted facing the pages upon which they are described:

ROSA, p. 17; MIGNONNE, p. 25; PRINCESS, p. 26; COOMASSIE, 2d, p. 30; LA FINANCIERE, p. 31; PRIMROSE, p. 33; MAUD, p. 35; BELLE DAME, p. 36; ANNA OF MOUNTAIN-SIDE, p. 53; NAMELESS GIRL, p. 63; SATIN, p. 69; BROWNIE, p. 71; BRUNETTE HAMMOND, p. 77; LILLE BONNE, p. 79; ANNIE GOLDDUST, p. 78.

☞ *The Frontispiece is a Group of Heifers sold in May, 1883; photographed in the autumn of 1882.*

EAR
I
MARK

Mouse (F. 1787), 11953. Chocolate-fawn, shading to gray; white tongue; mixed switch; white on brisket and belly. A superb, large, well-formed cow; level, deep, broad loined, wide hipped; limbs fine and straight; hide soft, thin, beautifully coated, unctuous; udder well formed.

History—Dropped spring of 1874. Bred by Philip M. le Neveu, St. Clements, Jersey. Qualified for admission to the Island Herdbook, April, 1878. Bought by E. Burnett, for T. A. Havemeyer, August, 1880, and imported to New York in steamship Marengo in September following.

Pedigree—Being of the *Foundation Stock* of the Island, *Mouse* has no recorded pedigree.

Progeny at Mountainside—Heifer, July 12, 1879, *Young Mouse*, No. 81, p. 88, by *Landseer* (P. 162).

Heifer, July 31, 1882, Mouse, 2d, 18447, by *Carlo* (P. 180 C.), 5559. Sold May 8th, 1883, at auction.

Rosa (F. 1896), 11954. [EAR MARK 2] Cream fawn, shading to gray, black tongue and switch. A superb cow, of large size and grand constitution. She has a beautiful head, set off with delicate, even, down-curving horns, large, dark-rimmed eyes and a broad muzzle; a clean neck, straight, broad back, low brisket, deep and capacious carcass, and low flanks; hips broad, rump long and wide, thighs long and flat, legs fine and straight; hide mellow, thin, and rich, with silky coat; udder capacious, with large teats, full, turgid milk-veins; escutcheon broad, well-expanded on thighs. She is from the herd of one of the best breeders on the Island, and is one of the handsomest and most useful cows of the herd.

History—Dropped in 1874. Bred by Albert le Gallais, La Moye, St. Brelade's, Jersey. Purchased of him by E. Burnett, for T. A. Havemeyer, August, 1880, and imported in steamer Marengo September 9th following.

Pedigree—*Rosa*, 11954; color, cream; breeder, A. le Gallais, St. Brelade's Parish, Jersey; dam, a cow bred by John le Brocq, St. Mary's Parish.

EAR

| 3 |

MARK

Amy la Grise, 11952. Solid light grayish fawn; full black points, tongue and switch. A medium-sized cow, "going all to milk," exhibiting a nearly perfect frame, with an enormous udder and large teats, tortuous milk veins, slender, incurving, waxy horns, silky coat and rich skin.

History—Dropped March, 1874. Bred by Wm. Amy, St. Peter's. Bought by E. Burnett, for T. A. Havemeyer, August, 1880, and imported in steamer Marengo in September following.

Pedigree

	COLOR.	BREEDER.	SIRE.	BREEDER.
Amy la Grise, 11952	Solid gr. fawn..	William Amy.	Bull on I. of Jersey.	William Amy
Dam, *Beauty*..........	————....	William Amy.	————..	————.. ....

The dam had repute as "a twenty-quart cow" on the Island, and the daughter is worthy of her descent.

Progeny at Mountainside—Heifer, 1878, *Northern Queen*, 11962, by *Northern Chief* (P. 137 H. C.).

Heifer, January 3, 1882, Amy la Grise, 2d, 18453, by *Carlo* (P. 180 C.), 5559. Sold in May, 1883.

Bull, November 18, 1882, Glory la Grise, 9136, by *Farmer's Glory* (F. 274 H. C.), 5196. Sold at the auction sale of May, 1883.

EAR MARK 4

Northern Queen, 11962. Gray fawn, with gray head, fringed ears, white fleck on brisket, spot on belly, white tongue, black switch. A large, handsome cow, very straight, broad, and level. She has a neat head, incurved, drooping horns, a clean neck, with low brisket, full crops, wide loin, deep and capacious carcass, broad hips, and long rump; legs straight and clean; hide mellow, loose, and unctuous, with a silky coat and creamy skin-color; udder large and soft, with good-sized teats, milk-veins prominent, and escutcheon running fully up and of good width.

History—Dropped February, 1878. Bred by Wm. Amy, St. Peter's, Jersey. Bought by E. Burnett for T. A. Havemeyer in August, 1880, and imported in September following in steamer Marengo.

Pedigree

	COLOR.	BREEDER.	SIRE.	COLOR.	SIRE'S BREEDER.
Northern Queen, 11962....	Gray fawn...	Wm. Amy.	Northern Chief (P. 137 H. C.)	Solid gray.	C. F. Dorey.
Dam, *Amy la Grise*, 11952.	Solid gray f'n.	Wm. Amy.	Bull owned by.............	——....	Wm. Amy.

2d Dam, Beauty, a 20-quart cow, bred by Wm. Amy.

AMY LA GRISE is a cow of extraordinary productiveness, richness, and beauty—No. 3 of this herd.

NORTHERN CHIEF is a highly commended bull and prize-winner, by Wm. Alexander's bull, *Gray Prince* (F. 168), and out of Mr. Dorey's famous *Queen of the North* (F. 1551 H. C.), she by Le Montais' bull, *Pansy* (F. 184 H. C.), and out of *Violet* (F. 997), bred by Francis le Brocq, a line of most excellent blood.

Progeny at Mountainside—Bull, October 27, 1881, Northern Carlo, 17112, by Hero (P. 126). Sold at auction in 1882.

EAR
5
MARK

Pansy (P. 330 H. C.), 11965. Fawn and white, shading to gray on the head; white spots and flecks on shoulders, forelegs, and flanks; white tongue and mixed switch; large, very level, square, and deep bodied; good head; light horn; hide mellow, loose, and unctuous, with a silky coat; udder capacious, with good-sized teats.

History—Dropped April 26, 1878. Bred by John Cooper, St. Ouen's, Jersey. She won *first*, "over all Jersey," as a yearling, in 1879, and *third*, in June, 1880. She was bought by E. Burnett, for Mr. Havemeyer, in August, and shipped to New York, in September, 1880, in steamship Marengo.

Pedigree

	COLOR.	BREEDER.	SIRE.
Pansy (P. 330 H. C.), 11965...	Fawn and white...	John Cooper...	*Browney* (P. 158 H. C.).
Dam, *Mergot* (F. 429)	Fawn and white...	John Cooper...	——————— ....

Progeny at Mountainside—Heifer, May 17, 1881. *Panftie,* 18442, by "*Avrill's Bull,*" on Island of Jersey.

Sold at auction, May 8, 1883.

EAR MARK | 6 |

Bijou le Cerf, 11964. Squirrel gray, with white spots on belly and fleck on flank, white tongue, black switch. A cow of large size, well formed, broad, and deep; neat head and horn, clean neck, thin withers, carcass capacious, loin and hips wide, rump long, thighs thin, tail and legs fine; excellent milk-veins; large udder and teats; escutcheon broad, and ascending fully up.

History—Dropped May, 1878. Bred by F. le Cerf, St. Ouen's Parish, Jersey. Won 3d prize at St. Ouen's Parochial Show in 1881. Bought by E. Burnett, for T. A. Havemeyer, in August, 1880, and imported in steamship Marengo in September following.

Pedigree

	COLOR.	BREEDER.	SIRE.	SIRE'S BREEDER.
Bijou le Cerf, 11964. Island name, *Bijou.*	Squirrel gray.	F. le Cerf, St. Ouen's, Jersey.	Prize bull.	Wm. Avrill.
Dam, *Polly*	——	F. le Cerf	——	——.

"Avrill's prize bull" was no doubt *Brownie* (P. 158), but of this we have no documentary evidence.

EAR
7
MARK

St. Ouen, 11963.   Solid dark fawn; black tongue and mixed switch.   A handsome cow, straight and deep, with thin shoulders, well-sprung ribs, capacious barrel; large, well-formed udder, with large teats; mellow and yellow skin.

History—Dropped April, 1878. Bred by Francis le Maistre, St. Ouen's, J. Bought by E. Burnett, for T. A. Havemeyer, in August, 1880, and shipped to New York by steamer Marengo in September.

Pedigree

	COLOR.	BREEDER.	SIRE.	COLOR.	SIRE'S BREEDER.
St. Ouen, 11963....	Solid fawn....	F. le Maistre..	*Browney* (P. 158 H. C.)..	Lt. brown...	Wm. Avrill..
Dam, *Princess*.. .	——........	Jas. le Maistre..	——.....	——......	——......

Progeny at Mountainside—Heifer, November 22, 1881,   St. Ouen, 2d, 18451, by *Carlo* (P. 180 C.), 5559. Sold May, 1883.

Heifer, October 12, 1882, St. Ouen, 3d 19698, by *Farmer's Pride*, 5560. Sold May, 1883.

EAR 8 MARK

Brunette le Breton, 11955. Solid dark mulberry brown, full black points, tongue, and switch. A cow of grand constitution and productiveness, of an exceedingly rich, deep color and good form. She is straight, wedge-shaped, deep, and broad in loin and hips; head long, with slender, drooping horns; neck clean and thin to withers, brisket low, body low and capacious, flanks deep, rump long and wide, thighs flat and long; legs straight and fine, and tail slender; hide mellow and unctuous, coat soft and silky, and the color of the skin in ears, tail tip, and on bare spots very rich; her udder is capacious, with long teats; milk-veins large and tortuous and escutcheon broad and high, of the flanderine type.

History—Dropped in 1875. Bred by Charles le Breton, St. Peter's, Jersey. Bought of Elie le Blancq, St. Ouen's, August 21, 1880, by E. Burnett, for T. A. Havemeyer, and shipped by steamer Marengo in September following.

Cocotte (P. 235 C.), 11958. Solid gray fawn, full black points, tongue and switch. A large, noble cow, closely approaching perfection in form. Her head is fine, dished, with incurving, drooping horns; neck and withers clean and thin, back level, and ribs well-sprung; brisket low, and line of the belly still lower and level; rump very long, thighs long and flat, legs clean and neat, tail slender; hide mellow and soft; skin color rich; udder capacious, with large teats and milk-veins, and a broad escutcheon, extending fully up.

RAR 9 MARK

History—Dropped February 25, 1876. Bred by Philip du Val, St. Peter's, Jersey. Bought by E. Burnett, for T. A. Havemeyer, August, 1880, and shipped to New York by steamer Marengo in September following. *Cocotte* won *first* over all Jersey at the May Show of the Royal Jersey Agricultural Society in 1880.

Pedigree

	COLOR.	BREEDER.	SIRE.	COLOR.	SIRE'S BREEDER.
Cocotte (P. 235 C.), 11958	Solid fawn.	Philip du Val	*Hero* (P. 90 H.C.)	Gray..	Pierre Poignaud.
Dam, *Belle* (F. 302).	Solid br'n..	J. du Val, St. Peter's, Jersey..	———	———	———

HERO won the *third* "Herdbook prize" in 1875, *first* St. Peter's in 1876, and *first* over all Jersey, and sweepstakes, the same year. He was by Welcome (F. 172 H. C.), and out of *Musique* (F. 1096 H. C.), both highly commended, Welcome a prize winner, and both animals of extraordinary excellence.

Progeny at Mountainside—Bull, November 12, 1881, Bruce of Mountainside, 7113, by *Carlo* (P. 180 H. C.), 5559. Sold in 1882.

EAR MARK: 10

Mignonne (P. 293), *11959*. Dark fawn, white fleck on brisket, black tongue and dark switch. A cow very nearly perfect in form, with a beautiful head, low set, incurving horns, straight-backed, broad and fine. She is of large size and great style, wedge-shaped, deep-bodied, low in flank and brisket, and has a grand udder, well quartered, with large teats; a beautiful, rich, and flexible hide, and first-class flanderine escutcheon.

History—Dropped February 8, 1877. Bred by Philip du Val, St. Peter's. Bought by E. Burnett, for T. A. Havemeyer, August, 1880, and imported to New York in steamer Marengo in September following.

Pedigree

	COLOR.	BREEDER.	SIRE.	COLOR.	SIRE'S BREEDER.
Mignonne (P. 293)	Dark fawn	Philip du Val	*Hero* (P. 90 H. C.)	Gray	P. Poignaud.
Dam, *Lily* (P. 70)	Gray	Philip du Val	*Pierrot* (F. 143)	Light gray.	Jas. Balliane.
2d Dam, *Belle* (F. 302)	Brown	J. du Val	—	—	—

HERO [Sire, *Welcome* (F. 172 H. C.); Dam, *Musique* (F. 1096 H. C.)] was a highly commended bull and winner of first prize and sweepstakes over the Island in 1876, whose sire and dam were both highly commended, and animals of great fame. *Mignonne's* son, *FARMER'S PRIDE*, 5560, by *Farmer's Glory*, probably the best son of that prolific sire, was imported with his dam, became an animal of extraordinary beauty and excellence, and was sold in May, 1883.

Progeny at Mountainside—Bull, July, 1880, *Farmer's Pride*, 5560, by *Farmer's Glory* (F. 274 H. C.), 5196. Bull, August 26, 1881, Mignonne's Duke, 7110, by *Carlo* (P. 180 C.), 5559. Sold in 1882.

EAR
11
MARK

Princess (F. 1646), 11957. Solid mulberry fawn, shading to gray and chocolate; full black points, tongue, and switch; a very neat, compact, fine-limbed, and well-formed cow; large-barreled, broad-loined, deep and square, with mellow, well-coated, rich hide; a dishing face; waxy horns; capacious, well-formed udder; large teats; escutcheon wide on thighs and high; large, tortuous milk veins.

History—Dropped in 1876. Bred by Ph. Bauche, St. Peter's, J. Bought by E. Burnett, for T. A. Havemeyer, in August, 1880, and imported in steamship Marengo, in September following.

Pedigree

	COLOR.	BREEDER.	SIRE.	COLOR.	SIRE'S BREEDER.
Princess (F. 1646), 11957..	Solid gr. br'n..	Philip Bauche..	*Gray Prince* (F. 168).	Silver gray..	W. Alexander.
Dam, unregistered cow....	———........	F. G. Dallain..	———............	———.....	———.......

Progeny at Mountainside—Heifer, March 15, 1881, *Glorinessa*, 18441, by *Farmer's Glory* (P. 274 H. C.), 5196. 11–1 of this Catalogue.

Heifer, March 20, 1882, Elsprita, 18459, by Duke of Darlington, 2460. Sold May, 1883.

EAR MARK | I I–I |

Glorinessa, 18441. Solid squirrel gray, white tongue, black switch; straight, broad, very deep and capacious, with fine head, rich horn, straight, deer-like legs, and fine tail; hide mellow and unctuous, with silky coat; skin color creamy and rich; udder and teats large and well formed.

History—Dropped March 15, 1881, at Mountainside. Bred by Philip Bauche, St. Peter's, Jersey. Bought, in her dam, by E. Burnett, for T. A. Havemeyer, and imported in steamship Marengo to New York in September, 1880.

Pedigree

	COLOR.	BREEDER.	SIRE.	COLOR.	SIRE'S BREEDER.
Glorinessa, 18441	Solid gray	Philip Bauche	*Farmer's Glory* (P. 274 H. C.)	Gray	Wm. Amy.
Dam, *Princess* (F. 1646)	Sol. gr. br'n	Philip Bauche	*Gray Prince* (F. 168)	Sil. gray.	W. Alexandre.
2d Dam, an Unregistered Cow.	—	F. G. Dalain.	—	—	—

PRINCESS is one of the choice cows of the herd—an abundant and persistent milker.

FARMER'S GLORY, grand himself, gets no inferior stock. A more promising combination could hardly be devised, and so far *Glorinessa* fulfils the highest anticipations.

EAR
	12	
MARK

Lady Arthur, 11966. Solid fawn, black tongue and switch.   A grand cow, level and

deep, broad in loin and pelvis, with straight, deer-like limbs, head well-poised, long face, slightly dished and slender in-curving horns; udder capacious, with medium-sized teats; escutcheon wide on thighs; hide mellow loose, and rich, with a soft silky coat.

History—Dropped in 1878.   Bred by John Arthur, St. Mary's.   Bought by E. Burnett, for T. A. Havemeyer, August, 1880, and imported in September of the same year in steamship Marengo.

As "Ladybird," her Island name, she won, as a yearling, the third prize "over all Jersey" in 1879.

Pedigree

FOLLOWING THE SIRE'S LINE.	COLOR.	BREEDER.	DAM.	COLOR.	DAM'S BREEDER.
Lady Arthur, 11966...........	Solid f'n..	John Arthur......	*Rozellette*...........	———	John Arthur.
Sire, *Royalist* (P. 139 H. C.)....	Dark gray.	J. P. Mourant....	*Regina* (F. 32 H. C.)..	Br'n & w...	J. P. Mourant.
2d Sire, *Duke* (P. 76 H. C.)... } *Sweepstakes Duke,* 1905 ..... }	Gray	Clement Lesbirel..	*Superb* (F. 353 H. C.).	Br'n & w ..	C. Lesbirel.
3d Sire, *Merry Boy* (P. 61 H. C.)	Gray.....	John Arthur......	*Eva* (F. 628).........	Br'n & w...	J. Arthur.
4th Sire, *Stockwell,* 2d (P. 24 H. C.)	Y. & w...	C. Godfrey.......	*Soucique* (F. 68)......	Y. & w.....	Ph. Godfrey.
5th Sire, *Noble* (F. 104 H. C.)..	Lt. br'n..	C. Pallot.........	———	———....	
6th Sire, *Sultan* (F. 58 H. C.)..	Lt. br'n..	C. Pallot.........	*Flower* (F. 53 H. C.).	Br. & w....	C. Lisbirel.
7th Sire, *Prince of Wales*......	——— ...	———	———	———	———

This is a long line of famous sires; all highly commended, and many of them first prize winners over the whole Island.

REGINA, the dam of Royalist, gave 18 lbs. Jersey weight of butter in seven days, equal to nearly 20 lbs. avoirdupois, and is said to have won every possible prize on the Island.

ROYALIST won first prize over the Island in 1876, and his sire, *Duke* (76), first and sweepstakes in 1875.   He (*Duke*) is the sire of GRAY KING (P. 160 H. C.), whose son, FARMER'S GLORY (F. 274 H. C.), gives renown to the entire line.   He is also the sire of VLRTUMNUS (P. 161 H. C.), hardly less famous.

Progeny at Mountainside—Heifer, March 3, 1882, Lady Arthur, 2d, 18458, by Duke of Darlington, 2460.
Sold in 1883.

EAR MARK 13

Arthur's Mistletoe, 11968. Solid gray fawn, very light on belly, black tongue and switch. A grand cow of large size and remarkably fine form—long, level, and deep. She has a beautiful head, finely carried, with light, drooping, incurving horns, and large, full eyes; neck and withers thin, back level and broad, brisket low, body capacious and level beneath, with low flanks; thighs long, rump high, limbs and tail fine; fore udder especially fine, teats large; hide showing excellent quality, very rich color in ears; escutcheon broad and high.

History—Dropped in 1879. Bred by John Arthur, St. Mary's, Jersey. Bought by E. Burnett, for T. A. Havemeyer, August, 1880, and imported in steamship Marengo in September of the same year.

Pedigree

FOLLOWING THE SIRE'S LINE.		COLOR.	BREEDER.	DAM.	COLOR.	DAM'S BREEDER.
Arthur's Mistletoe.. Island name, *Mistletoe.*	11968	Solid gr. fawn	John Arthur	*Rosebud*	—	John Arthur.
Sire, *Tormentor* (F. 259 H. C.)		Gray	John Arthur	*Angela* (F. 1606)	Brown.	P. de la Cour.
2d Sire, *Khedive* (P. 103 H. C.)		Light brown	Ph. Godeaux	*Coomassie* (F. 1442 H.C.)	Brown.	C. F. Dorey.
3d Sire, *Leo* (F. 198 H. C.)		Gray	Josué le Gros	—	—	—

ROSEBUD is one of Mr. Arthur's famous Rose family, remarkable for her great "richness."

TORMENTOR and *KHEDIVE* are famous, not alone by reason of their descent from *COOMASSIE*, but for the extraordinary butter records of their progeny.

EAR MARK | 14 |

Coomassie, 2d, 11969. Solid fawn, except star in forehead, black tongue and switch. A large cow, and one of remarkable presence and beauty, having a level back, with great breadth of loin and hips, a deep and capacious carcass, well-ribbed back; head broad, dished, with full eyes, well-shaped incurving horns; neck and withers clean; brisket low, and belly-line level; milk veins full and tortuous; udder perfect in form, with large, well-placed teats; escutcheon selvage of the first order, with very deep thigh ovals; hide mellow, coat furry and soft, with rich orange skin color.

History—Dropped March 22, 1879. Bred by C. F. Dorey, Trinity Parish, Jersey. Bought by E. Burnett, for T. A. Havemeyer, August 21, 1880, and imported in steamship Marengo to New York in September following. The only living daughter of her famous dam.

Pedigree

	COLOR.	BREEDER.	SIRE.	COLOR.	SIRE'S BREEDER.
Coomassie, 2d, 11969	Fawn....	C. F. Dorey.	*Guy Fawkes* (F. 251 H. C.)	Sol. lt. gr.	Philip Godeaux.
Dam, *Coomassie* (F. 1442 H. C.)	Brown...	C. F. Dorey.	*Neptune* (P. 14 C.)	Dark br'n.	D. Blampied.
2d Dam, *Jersey Pride* (F. 1716 H. C.).	Br'n & w..	C. F. Dorey.	*Sans Peur* (P. 2 H. C.)....	Br'n & w.	P. Nicolle.

3d Dam, Aged Cow of Mr. Dorey's not presented for qualification in the Herdbook.

COOMASSIE is one of the most remarkable cows that ever left the Island, having won the *first* prize over all Jersey five consecutive years, 1876 to 1880 inclusive, and in her descendants she is still more famous. So remarkably are her qualities transmitted, particularly her beautiful udder and her butter qualities, that "the Coomassies" are easily recognizable to the fourth generation, and really form a family by themselves. Her reported butter yield is 16 lbs. 11 oz. in 7 days.

GUY FAWKES is her grandson, by *Koffee* (F. 233 H. C.), and out of *Angelica*, a daughter, by *Orange Skin*, of the famous *Garenne* (F. 1575 H. C.). Thus we have in COOMASSIE, 2d, not only an in-bred daughter and great-granddaughter of the famous cow, but the *Garenne* cross so highly prized in this combination.

The daughters of *Guy Fawkes* are famous as butter producers. *Island Star* is credited with 15 lbs. 8 oz. as a 3-year-old, Anntybel with 14 lbs. 9 oz. at same age, and *Queen of Ashantee*, six months after calving, as a 2-year-old, with 10 lbs. 14 oz.

Progeny at Mountainside—Bull, October 1, 1881, Lord of Mountainside, 7111, by *Carlo* (P. 180 C.). Sold at auction in May, 1882.

EAR 15 MARK *La Financiere, 11970.* Solid silver-gray fawn; gray head; black tongue and switch. And extraordinarily level, deep, broad-loined, fine-boned cow, wedge-shaped and square, with a beautiful head, light "wild" horn, thin neck and withers; hide soft, with silky and unctuous coat.

History—Dropped December, 1878. Bred by Chas. Norman, Trinity Parish, Jersey. Bought by E. Burnett for T. A. Havemeyer, August, 1880, and imported in steamer Marengo to New York, in September of the same year.

Pedigree

FOLLOWING THE LINE OF SIRES.	COLOR.	BREEDER.	DAM.	COLOR.	SIRE'S BREEDER.
La Financiere, 11970	Solid gr. f'n	Chas. Norman	*La Picote*	——	M. Picot.
Sire, *Grey King* (P. 169 H. C)	Silver gray.	Wm. Alexandre	*Lily Gray* (F. 770)	Silver gr	Wm. Alexandre.
2d Sire { *Duke* (P. 76 H. C.) / *Sweepstakes Duke,* 1905 }	Gray	Clement Lesbirel.	*Superb* (F.353 H.C.)	Br. & w.	C. Lesbirel.
3d Sire, *Merry Boy* (P. 61 H. C.)	Gray	John Arthur	*Eva* (F. 628)	Br. & w.	J. Arthur.
4th Sire, *Stockwell,* 2d (P. 24 H. C.)	Yel. & w.	C. Godfrey	*Soucique* (F. 68)	Y. & w.	Ph. Godfrey.
5th Sire, *Noble* (F. 104 H. C.)	Lt. brown	C. Pallot	——	——	——
6th Sire, *Sultan* (F. 58 H. C.)	Lt. brown	C. Pallot	*Flower* (F. 53 H. C.)	Br. & w.	C. Lesbirel.
7th Sire, *Prince of Wales*	——	——	——	——	——

A long line of famous, highly commended sires, nearly all winners of the highest prizes.

GREY KING (P. 169) is the sire of *Farmer's Glory* (P. 274 H. C.). I. of this Catalogue. In connection with his pedigree there will be found much of interest concerning him and his get.

DUKE (P. 76 H. C.) is the sire of *Vertumnus* (P. 161 H. C.), whose daughters, *Punchinello,* 11875, *Day Dream* (P. 339 H. C.), *La Rouge,* 12405, *Tidy,* 2d, 10925, the dam of *Tadcaster,* 5150, grace the choice herds of Messrs. Burnham and Wing.

Progeny at Mountainside—Heifer, February 19, 1882, La Financiere, 2d, 18456, by Duke of Darlington, 2460. Sold May, 1883.

EAR 16 MARK

Falladot, 11967. Fawn, with white spots on belly; black tongue and switch. A medium-sized, well-shaped, very deep-bodied cow, with a neat head and incurving horns; back broad, brisket low, and belly level with it; barrel capacious, hips long and wide, rump slightly drooping; legs straight, and fine in bone; tail long and fine; hide thin and mellow, coat silky, color in skin and ears creamy; udder of excellent form, and capacious; teats of fair size; escutcheon well out on thighs, and running fully up.

History—Dropped February 17, 1879. Bred by Thos. Falla, Jr., St. John's Parish, Jersey. Bought by E. Burnett, for T. A. Havemeyer, in August, 1880, and imported in steamer Marengo to New York in September following.

Pedigree

	COLOR.	BREEDER.	DAM.	COLOR.	DAM'S BREEDER.
Falladot, 11967............	Fawn & white..	T. Falla, Jr..	*Cherry* (F. 1140)	Light red..	Thos. Falla.
Sire, *Carlo* (P. 180), 5559...	Solid orange f'n.	T. Falla, Jr..	*Pretty Maid* (F. 1493).	Brown	T. Falla, Jr.
2d Sire, *Hero* (P. 126 H. C.)	Gray..........	T. Falla, Jr..	*Cowslip* (P. 24)	Brown	Thos. Falla.
3d Sire, *Dick* (F. 171 H. C.)	Light brown ...	T. Falla, Jr..	*Cherry* (F. 1140)	Light red..	Thos. Falla.
4th Sire, *Yankee* (P. 27 H.C.)	Light brown ...	E. Gibaut....	*Georgette* (F. 309)......	Red	E. Gibaut.
5th Sire, *Paddy* (F. 97)....	Light brown ...	E. Gibaut....	———	———	———

CHERRY, her dam, is also the dam of *Dick,* grandsire to *Carlo.* *YANKEE* also occurs twice in *Carlo's* pedigree. *PRETTY MAID,* her dam, *Tinker,* and granddam, *White Rose,* all owned and bred by Mr. Falla, or his father, were a breed of milkers having great fame for their butter product, a fact which led to the selection of *CARLO* by Mr. Burnett for his position in this herd.

CARLO (P. 180). See full pedigree upon page 10 of this Catalogue.

YANKEE (P. 27 H. C.) won *first* over all Jersey in 1871, *first* Herdbook Society's prize, and the silver medal of the Channel Island Exhibition, the same year.

EAR MARK 17 *Primrose* (P. 244), 11956. Brownish fawn, with white under the belly; white tongue and mixed switch; level and broad, wide rumped, square and deep, with thin withers and neck; fine head and horns; udder and teats large; milk-veins full and tortuous; hide mellow and rich, with furry coat, and rich skin color.

History—Dropped February 3, 1876. Bred by Thos. Falla, Jr., St. John's Parish. Bought by E. Burnett, for T. A. Havemeyer, August, 1880, and imported to New York in steamship Marengo in September following.

Pedigree

	COLOR.	BREEDER.	SIRE.	COLOR.	SIRE'S BREEDER.
Primrose (P. 244), 11956..	Light yellow..	Thos Falla, Jr..	Tom (P. 77 H. C.)..	Dark gray...	Philip du Val.
Dam, *Cherry* (F. 1140)...	Yellow fawn..	Thomas Falla..	——............	——......	——......

TOM is a highly commended bull, and a prize winner, by *Welcome* (F. 172 H. C.), a first prize winner of St. Peter's show in 1873 and in 1874, and one whose name occurs in many of the most famous Island pedigrees.

Progeny at Mountainside—Heifer, January 3, 1882, Carlo's Primrose, 18454, by *Carlo* (P. 180 H. C.), 5559. Sold May, 1883.

Sultane 4th (P. 276 C.), 11960.

EAR MARK 19

Sultane 4th (P. 276 C.), 11960. Dark brown, nearly black, small spot of white on each flank, black tongue, and mixed switch. A medium-sized cow, of excellent shape, level, broad, and deep; head delicate, with neat, incurving horns; neck thin and clean to withers, back straight, brisket low, belly line very low; flanks low, thighs long and flat, hips wide, rump long and wide, legs straight and clean, tail fine; hide soft and coat silky, with creamy skin color; escutcheon large, broad, and running fully up; udder well-shaped and capacious.

History—Dropped April 25, 1877. Bred by John P. Marrett, St. Saviour's Parish, Jersey. Bought by E. Burnett, for T. A. Havemeyer, August, 1880, and imported in steamship Marengo in September of the same year.

Pedigree

	COLOR.	BREEDER.	SIRE.	COLOR.	SIRE'S BREEDER.
Sultane, 4th, (P. 276 C.), 11960.......	Dark brown.	J. P. Marrett.	*Khedive* (P. 103 H. C.)	Lt.br'n.	C. F. Dorey.
Dam, *Sultane* (P. 7 H. C.).............	Black.......	J. P. Marrett.	*Sultan* (F. 58 H. C.)	Lt.br'n.	C. Pallot.
2d Dam, *Longueville Queen* (F. 272 H.C.)	D'k red & w.	J. P. Marrett.	—————........ ...	———.	

A rare combination of famous blood.

LONGUEVILLE QUEEN, the dam of *Sultane*, is a cow long to be remembered by visitors to the Island as a typical Jersey cow of great beauty and productiveness, as well as for being the "queen" of one of the best herds of the Island.

SULTANE, the most famous daughter of *Sultan*, 58, won *second* over all Jersey in 1870, and first St. Peter's the same year; *first* Herdbook prize, and first at the Channel Island Exhibition in 1871.

KHEDIVE is one of the best sons of the famous *COOMASSIE*. He won *second* over all Jersey, *first* Herdbook, and *first* Trinity Parochial in 1876.

SULTAN won *first* over all Jersey in 1867 and 1868, and *first* at the Royal, in England, in the latter year.

Progeny at Mountainside—Bull, August 12, 1881, Sultan Carlo, 7108, by *Carlo* (P. 180 C.), 5559. Sold at auction in 1882.

Maud (*F. 2172*), *11961*. Light fawn, with white between forelegs, and flecks on shoulder and flank; white tongue and black switch; above medium size; level, deep-bodied, well-formed, and handsome, with waxy, blue-tipped horns; dark-fringed ears; large, well-quartered udder; large teats and milk-veins; a uniform, moderately-wide escutcheon, running fully up.

EAR MARK [20]

History—Dropped in 1877. Bred by Francis le Brocq, St. Peter's. Bought by E. Burnett, for T. A. Havemeyer, August, 1880, and shipped to New York, in steamer Marengo in September following.

Pedigree—Being of the *Foundation Stock* of the Island, Maud has no recorded pedigree.

Progeny at Mountainside—Heifer. March 25, 1881, *Carlo's Maud*, 18452, by *Carlo* (P. 180 C.), 5559. Sold May, 1883.

EAR
21
MARK

Belle Dame (F. 1934 H. C.), 11951. Solid dark mulberry fawn, with black tongue and switch. A wonderfully well-formed cow, showing strong constitution, great digestive capacity; she is level, broad, deep, and wide in the hindquarters; fine boned; light limbed, with a large, well-formed udder; good teats; long meandering milk-veins; a thin, fine-haired, mellow skin, and a broad and high escutcheon, with extra ovals and curls enough to please any connoisseur in such matters.

History—Dropped October, 1872. Bred by Francis le Brocq, St. Peter's. Bought by E. Burnett, for T. A. Havemeyer, August, 1880, and imported, in steamship Marrengo, in September of that year.

Pedigree—Being of the *Foundation Stock* of the Island, she has no recorded pedigree, but, at the examination for admission to the herd book, was highly commended.

Progeny at Mountainside—Bull, March 6, 1882, Pride of Mountainside, 7118, by Duke of Darlington, 2460. Retained for use in the Herd—No. 21–2, page 12.

EAR |22| **MARK**

Lindora, 5361. Gray, shaded with fawn, with white on withers and belly, and flecks on flanks; white tongue and mixed switch. A large, well-shaped cow, with a neat head and fine horns; back high at the hips, flanks low, brisket and belly-line very low, thighs full, rump broad; legs straight, and fine in bone; tail fine; hide mellow and loose, coat silky and unctuous, and skin color rich; escutcheon, broad selvage; udder of good size and well shaped.

History—Dropped May 1, 1874. Bred by B. Kittredge, Peekskill. Bought at auction, May, 1881. Won 2d prize at the show of the New York State Agricultural Society, in 1876.

Pedigree

	COLOR.	BREEDER.	SIRE.	COLOR.	SIRE'S BREEDER.
Lindora, 5361	Gray	B. Kittredge	*St. Martin*, 1482	Solid gray	B. Kittredge, impr.
Dam, Lucilla W., 5357	Solid gray	B. Kittredge	*Barney*, 1491	Solid brown	J. C. Godfray.
2d Dam, *Lucy K.*, 5262	Gray	P. Morant	—	—	—

EAR
23
MARK

Crescent of Maple Shade, 10927. Brown, shading to gray, nearly solid; a crescent of white in forehead and white spot back of brisket; black switch and black-tipped tongue; a neat, well-formed cow, with a well-shaped udder and teats; good milk veins; escutcheon extending far out on thighs; soft, mellow hide and fine coat.

History—Dropped June 23, 1878. Bred by Amice Alexandre, St. Peter's, Jersey. Imported by John D. Wing, August, 1880, in steamship City of London to New York. Bought at auction, May, 1881.

Pedigree

	COLOR.	BREEDER.	SIRE.	COLOR.	BREEDER.
Crescent of Maple Shade, 10927..	Brown...	A. Alexandre..	*Browney* (P. 158 H. C.)..	Light br'n.	Wm. Avril.
Dam, *Little Browney* (P. 29)....	Brown...	A. Alexandre..	*Brown Prince* (F. 85 H.C.)	Brown....	Elias Nicole.
2d Dam, *La Hoguette* (F. 167)...	Brown...	A. Alexandre..	———	———....	———

Progeny at Mountainside—Heifer, Oct. 7, 1881, Crescepola, 18450, by Polonius, 2513. Sold May, 1883.

EAR MARK `|24|`

Paraphrase, 10254.
Gray fawn, with white on belly, white tongue and switch. A rather tall cow, straight-backed, with wide hips and rump; well-sprung ribs, thin neck, neat, well-set head, with light horns; well-shaped, good-sized udder and teats.

History—Dropped April 19, 1879. Bred by Thos. J. Hand. Transferred to John D. Wing, in 1880, and bought at auction at the sale of the Mapleshade herd in May, 1881.

Pedigree

	COLOR.	BREEDER.	SIRE.	COLOR.	SIRE'S BREEDER.
Paraphrase, 10254.............	Gray fawn....	T. J. Hand..	Medalist, 2122	Solid dark gray.	C. M. Beach.
Dam, Belle of Poquonnock, 1874.	S'd Smoky f'n.	William Clift.	Ossipe, 697...	Solid brn. & tan	Jas. P. Swain, Sr.
2d Dam, *Sallie Bunker*, 1426....	Gray fawn....	M. Alexandre, St. Ouen's, Jersey.			

MEDALIST combines the famous blood of *Splendid*, 2, Albert, 44, Lord Ogden, 69, and *Pierrot*, 636.

Through her dam, Paraphrase is connected with the Bird, Maitland, and Cameron importations, from each of which many excellent animals have sprung.

Progeny at Mountainside—Bull, Nov. 17, 1881, Glory of Mountainside, 7114; by Dido's Duke, 4678. Sold in 1882.

Bull, September 26, 1882, Translation, 9053, by Black Prince of Hanover, 2873. Sold May, 1883.

EAR
25
MARK

Troth Plight, 10258.

Solid mulberry fawn, shading to black on head and neck; black tongue and switch; level, deep in body and flank; neat head, with light, up-set horns, fine bone; hide very loose and mellow, with rich color in ears; udder large, with irregular good-sized teats; escutcheon well out on thighs.

History—Dropped June 16, 1879. Bred by Thos. J. Hand. Transferred to John D. Wing in 1880, and sold with the Mapleshade herd, May, 1881.

Pedigree

	COLOR.	BREEDER.	SIRE.	COLOR.	SIRE'S BREEDER.
Troth Plight, 10258.....	Solid fawn......	T. J. Hand......	Medalist, 2122...	Solid dark gr..	C. M. Beach.
Dam, Troth, 6139......	Solid fawn......	T. J. Hand......	Hornbeam, 2123..	Solid dark gr..	T. J. Hand.
2d Dam, *Blondette*, 1817.	Solid lemon f'n..	Mr. Noel, Jersey.	——..........	-——.......	——

Through MEDALIST, Troth Plight descends from the famous New England bulls, *Splendid*, 2, Albert, 44, Lord Ogden, 69, and *Pierrott*, 636; while on the dam's side, Hornbeam, 2123, is, through his sire, Marius, 760 (the sire of Signal, 1170), double *Lady Mary* (one of the noblest cows ever imported), combined with the choice blood of *Southampton*, 117, and *Emblem*, 90, in Emily Hampton, his dam. A richer combination it would be hard to devise. Mr. Hand's herd has twice won the great Gold Herd Medal of the New York State Agricultural Society, besides winning most of the high prizes whenever exhibited. TROTH, 6139, made over 16 lbs. of butter in seven days, for her owner, Chas. E. Hand.

Progeny at Mountainside—Heifer, November 4, 1881, Troth Plight, 2d, 18449, by Dido's Duke, 4678. Sold May, 1883.

Witch of the Maples, 10093. Solid fawn, with fringed ears, black tongue and switch. A large cow, with cylindrical body, level, with drooping rump; head light, face dished, horns light and up-set, back broad in loin, brisket low, legs strong and straight; hide mellow, furry, and soft, skin-color, creamy; she has a good coupé udder, with two rudimentary teats and a flanderine escutcheon.

EAR 27 MARK

History—Dropped February 5, 1880. Bred by John D. Wing. Bought at the sale of the Mapleshade Herd May, 1881.

Pedigree

	COLOR.	BREEDER.	SIRE.	COLOR.	SIRE'S BREEDER.
Witch of the Maples, 10093....	Solid fawn.	John D. Wing..	Niobe's Duke, 2364..	Solid.....	S. J. Sharpless
Dam, Witch Hazel, 3d, 4875....	Gray fawn.	T. J. Hand....	Marius, 760.........	Solid gray.	W. H. Scheifflin
2d Dam, Witch Hazel, 1360.....	Fawn & w..	T. J. Hand....	*Southampton*, 117....	Or. brown.	Ph. Gaudin.
3d Dam, *Hazel*, 91.............	Reddish f'n.	Ph. Gaudin.....	*Clement* (F. 61 H.C.)	D'k f'n & w.	E. Gibaut.
4th Dam, *Lady Bird* (F. 1551 H.C.)	Gray & w..	Ph. Gaudin—J. H. McHenry, impr.			

Empresa, 4th, 12278.

BAR 28 MARK

Solid fawn shading to gray; black tongue and switch; a medium-sized cow, very level and square, with broad loin and rump; excellent udder, especially good, forward medium-sized teats, full milk veins; hide mellow and loose, with unctuous, silky coat.

History—Dropped April 27, 1880. Bred by John D. Wing, and bought by Mr. Havemeyer, at the sale of Mr. Wing's herd in 1881.

Pedigree

	COLOR.	BREEDER.	SIRE.	COLOR.	SIRE'S BREEDER.
Empresa, 4th, 12278...	Solid fawn ..	John D. Wing....	Babylon, 4723....	Solid dark fawn..	John D. Wing.
Dam, Empresa, 2790...	Solid fawn ..	T. J. Hand	Marius, 750......	Solid gray.......	T. J. Hand.
2d Dam, *Emblem*, 90..	Gray & white.	E. Gibant........	——............	——..........	——..........

BABYLON is of Faile's famous "Edith" family, and combines the blood of Saturn, 94 (the sire of Alphea), Swain's Lord Nelson, and Wing's Devil's Hoof; while Marius, 750 (the sire of Signal, 1170), by *Willie Boy*, 434, out of his dam, *Lady Mary*, 1148, is an example of close in-breeding productive of the very best results in impressing the admirable qualities of *Lady Mary* upon a large family of excellent cows. Among Signal's daughters we find Tenella, 6712, reputed to have made 22 lbs. 1½ ozs. of butter in 7 days; Croton Maid, 1505, credited with 21 lbs. 11½ ozs.; Valhalla, 5500, with 16 lbs.; and Edwina, 6713, with 15 lbs. 13 ozs.

Progeny at Mountainside—Bull, April 7, 1882, Impress, 9044, by *Farmer's Pride*, 5560. Sold at auction, May, 1883.

[EAR 29 MARK] *Parasol, 12658.* Solid chocolate-fawn, shading to black on face and legs; black tongue and switch. A neat, pretty creature, with beautiful light head and horns, straight back, wide loin, hips, and rump; tail and limbs fine; hide mellow and skin unctuous, with silky coat and rich color; udder well-formed, with teats above medium size and squarely placed.

History—Dropped, March 3, 1880. Bred by J. T. Michel, St. Peter's, Jersey. Imported by John D. Wing in steamship Marengo, to New York, January, 1881, and bought at auction in May following.

Pedigree

FOLLOWING THE LINE OF SIRES.	COLOR.	BREEDER.	DAM.	COLOR.	DAM'S BREEDER.
Parasol, 12658	Solid fawn.	J. T. Michel	*Faith* (F. 2009)	Lt. br'n	J. T. Michel.
Sire, *Farmer's Glory* (F. 274 H. C.)	Gray	F. Le Brocq	*Bonheur* (F. 1651 H. C.)	Lt. br'n	F. Becquet.
2d Sire, *Grey King* (P. 169 H. C.)	Silver gray.	Wm. Alexandre	*Lily Grey* (F. 770)	Slvr. gr.	Wm. Alexandre.
3d Sire, *Duke* (P. 76 H. C.) / *Sweepstakes Duke*, 1905	Gray	Clement Lesbirel	*Superb* (F. 353 H. C.)	B. & w.	C. Lesbirel.
4th Sire, *Merry Boy* (P. 61 H. C.)	Gray	John Arthur	*Eva* (F. 628)	B. & w.	J. Arthur.
5th Sire, *Stockwell* (P. 24 H. C.)	Y. & w.	C. Godfrey	*Soucique* (F. 68)	Y. & w.	Ph. Godfrey.
6th Sire, *Noble* (F. 104 H. C.)	Lt. brown.	C. Pallot	—	—	.
7th Sire, *Sultan* (F. 58 H. C.)	Lt. brown.	C. Pallot	*Flower* (F. 53 H. C.)	B. & w.	C. Lesbirel.
8th Sire, *Prince of Wales*	—	—	—	—	.

This long line of illustrious sires stands unequalled among Jersey pedigrees. They are all highly commended, nearly all are winners of first prizes at the shows of the Royal Jersey Agricultural Society, and to catalogue their achievements and the butter records of their immediate descendants would fill pages. The names of the animals recur frequently in this Catalogue, where additional information will be found.

Progeny at Mountainside—Bull, April 5, 1882, Parachute, 9043, by *Farmer's Pride*, 5560. Sold May, 1883.

Fauvette (F. 1801), 15021. Solid fawn, black tongue, and switch with few white hairs; a very level, low-bodied, wide-hipped, broad-loined cow, of excellent constitution, having a good head with strong, incurving, waxy horns; a large udder, with good-sized teats, and strong, tortuous milk veins; a mellow hide, with very silky, unctuous coat.

History—Dropped in 1874. Bred by Philip Nicole, Grouville, J. Qualified for Herdbook, April 25, 1878. Imported, by E. P. P. Fowler, to Philadelphia, in steamship Lord Clive, in January, and sold at auction by Herkness, in Philadelphia, May 12, 1881.

Pedigree—Being of the *Foundation Stock* of the Island, *Fauvette* has no recorded pedigree.

Progeny at Mountainside—Bull, April 14, 1882, Grilpolo, 9045, by Black Prince of Hanover, 2873. Sold at auction, May, 1883.

☐ 31 · MARK *Gambette, 2d, 15541.* Squirrel gray; white on belly and legs; white tongue and switch. Very level, broad and deep, long-rumped, wide-hipped, and square; head very fine, with exceedingly short incurving horns, and large full eyes; udder large, especially good forward, with good-sized teats, and prominent meandering milk veins.

History—Dropped spring of 1877. Bred by Philip Labey, Grouville. Imported by E. P. P. Fowler, in steamer Lord Clive, to Philadelphia, January, 1881. Sold at auction by Herkness, in Philadelphia, May, 1881.

Pedigree

	COLOR.	BREEDER.
Gambette, 2d, 15541	Squirrel gray	Philip Labey, Grouville, Jersey.
Dam, *Gambette*	——	T. Gaudin, St. Martin's, Jersey.

Progeny at Mountainside—Heifer, July 2, 1882, Whipped Cream, 19528, by Gilderoy, 2107. Sold May, 1883.

EAR MARK 32

St. Martainaise, 15543. Reddish fawn, with white flecks on belly, flanks, and right of rump; white tongue and black switch. A fine, large cow, level, deep, and cylindrical, with a handsome head, well carried, and a light horn, a straight back, wide at loin and hips, low brisket, broad chest, low flank, capacious carcass; rump long and wide, thighs long and flat, legs strong and straight, tail long, with heavy switch; hide exceedingly soft and mellow, with silky coat and rich skin color; udder large, with well-shaped, large teats, two rudimentary ones; milk veins full, escutcheon broad, "*pot de vin.*"

History—Dropped April, 1878. Bred by John Priaulx, Jersey. Imported by E. P. P. Fowler, January, 1881, in steamer British Queen via Liverpool to New York. Sold at auction, May 12th, by Herkness, in Philadelphia.

Pedigree

FOLLOWING THE LINE OF SIRES.	COLOR.	BREEDER.	DAM.	COLOR.	DAM'S BREEDER.
St. Martainaise, 15543..... Island name, *St. Martinaise, 2d.*	Reddish fawn.....	John Priaulx..	*St. Martinaise*..	——....	John Priaulx.
Sire, *Browny* (P. 158 II.C.)...	Solid light brown..	Wm. Avril ...	*Fairy* (F. 964)..	Light red..	Nicholas Arthur.
2d Sire, *Tom* (P. 77 II. C.)...	Solid dark gray...	P. du Val	*Belle* (F. 302)..	Brown....	J. du Val.
3d Sire, *Welcome* (F. 172 H. C.)	Gray	A. Poignand..	——.........	——....	——

☐ 33 ☐
BAR
MALE

Nelly of Mountainside, 15545. Solid gray fawn, with black tongue and dark brown switch. A very neat little cow, slender and deer-like, with a beautiful head, full eyes, long, dished face, small up-set horns; back level and broad, with thin withers; neck clean and thin; body cylindrical; thighs long and flat, rump wide and long; legs clean, straight, and fine; hide mellow, coat silky, and skin color creamy; udder well shaped, with good capacity and long teats; escutcheon of moderate extent (flanderine).

History—Dropped February, 1879. Bred by George Mauger, St. John's Parish, Jersey. Bought by E. P. P. Fowler, for A. M. Herkness & Co., Philadelphia, January, 1881. Shipped to New York by steamer Lord Clive from Liverpool the same month, and sold to Mr. Havemeyer, at the Herkness Auction Sale, in May of the same year.

Pedigree—Sire, bull of John Arthur, of St. Mary's, for 1878. Dam, cow owned by Mr. Mauger.

This pedigree, though in bad form, considered with the fact that *Nelly* was imported in calf to *Grey of the West* (F. 317 H. C.), one of John Arthur's best bulls, and winner of the 3d prize at the London Dairy Show in 1880, indicates that her breeder has habitually sent his cows for service to Mr. Arthur's best bulls, than which there have been no better.

EAR | **34** | **MARK**

Romulus Lively, 15024. Fawn and white; white tongue and mixed switch. A neat, well-formed cow, above medium size, level, broad, and deep; wide hips and rump; hide soft and mellow, with rich skin color; udder large, especially good forward, with large milk veins; escutcheon well out on thighs.

History—Dropped February 19, 1879. Bred by John du Val, St. Peter's Parish, Jersey. Bought by E. P. P. Fowler, January, 1881, imported in steamship Lord Clive to Philadelphia, the same month, and bought by Mr. Havemeyer, at the Herkness Sale, in May following.

Pedigree

FOLLOWING THE LINE OF SIRES.	COLOR.	BREEDER.	DAM.	COLOR.	DAM'S BREEDER.
Romulus Lively, 15024	F'n & w..	John du Val.....	*Daisy* (205).............	Yel. f'n..	J. du Val, Jr.
Sire, *Romulus* (P. 181 H. C.)..	Gray	F. le Brocq, Jr...	*Stella* (F. 705).........	Brown ..	John le Brocq.
2d Sire, *Hero* (P. 90 H. C.)....	Dark f'n.	Pierre Poignaud.	*Musique* (F. 1096 H. C.).	Dark f'n.	P. Poignaud.
3d Sire, *Welcome* (F. 172 H. C.)	Gray	A. Poignaud.....	———.................	——— ..	———

Progeny at Mountainside—Bull, March 6, 1882, Lively Carlo, 9041, by *Carlo* (P. 180 H. C.), 5559. **Sold** May, 1883.

EAR [35] **MARK**

Redbreast (P. 175), 15022. Fawn, shading to gray on head, red poll, white triangle in forehead, white brisket and belly, ring on tail, white tongue, black switch. A magnificent cow, above medium size, very straight, broad, and deep. She has a neat head, nearly black, with a fine eye and rich, waxy horn, brisket exceedingly low, chest wide, thick through the heart, long and wide rump; hide thin, loose, and mellow, with a short, silky coat and very rich yellow skin color; udder mellow and very capacious, with good-sized teats; milk veins full and tortuous; escutcheon broadly expanded on thighs (selvage).

History—Dropped April 13, 1875. Bred by Philip Binet, Trinity Parish, Jersey. Sold in December, 1880, to E. P. Fowler, shipped by steamship Lord Clive in January, and bought by Mr. Havemeyer, at the Herkness Sale, in May, 1881.

Pedigree

FOLLOWING THE SIRE'S LINE.	COLOR.	BREEDER.	DAM.	COLOR.	DAM'S BREEDER.
Redbreast (P. 175)	Fawn	Philip Binet	*Butter Cup* (F. 498)	Br'n & w.	P. Binet.
Sire, *General Don* (P. 60 H. C.)	Br'n & w.	J. P. Nicolle	*Milk Maid*, 3d (P. 4)	Br'n & w.	J. P. Nicolle.
2d Sire, *Emperor* (P. 16 H. C.)	Br'n & w.	C. Lesbirel	*Superb* (F. 353 H. C.)	Br'n & w.	C. Lesbirel.
3d Sire, *Toby* (F. 108)	Dark gray.	C. Lesbirel	——	——	——

REDBREAST draws her blood from some of the best herds in Jersey. Lesbirel, Nicolle, and Binet are all breeders of distinction.

SUPERB, Emperor's dam, was also the dam of *Duke*, 76, imported as *Sweepstakes Duke* by Moses Ellis. She is by the famous bull, *Prince of Wales*.

Progeny at Mountainside—Bull, January 16, 1882, *Darlo's Glory*, 7116, by *Darlo*. Sold in 1882.

EAR | 36 | **MARK**

Bergerelia, 15546. Dark mulberry fawn, with gray-tipped hairs, black head, white on belly and flanks, black tongue and switch. A well-shaped, deep-bodied, wide-hipped, broad-loined cow of strong constitution ; neat head and horns ; large udder, with grand milk veins and a wide escutcheon ; hide mellow and loose ; coat, soft and furry, with rich skin color.

History—Dropped December, 1878. Bred by John du Val, St. Peter's, Jersey. Sold to E. P. P. Fowler, and shipped by steamer Lord Clive, in January, 1881. Bought by Mr. Havemeyer, at Herkness Sale, in May of the same year.

Pedigree

FOLLOWING THE LINE OF SIRES.	COLOR.	BREEDER.	DAM.	COLOR.	DAM'S BREEDER.
Bergerelia, 15546............ (Island name, *Bergere.*)	Mul. f'n.	John du Val.....	Unnamed cow..........	——— ..	John du Val.
Sire, *Romulus* (P. 181 H. C.)..	Gray	F. le Brocq, Jr..	*Stella* (F. 705)...........	Brown ..	John le Brocq.
2d Sire, *Hero* (P. 90 H. C.)....	Dark f'n.	Pierre Poignaud.	*Musique* (F. 1096 H. C.).	Dark f'n.	P. Poignaud.
3d Sire, *Welcome* (F. 172 H. C.)	Gray	A. Poignaud....	———	——— ..	———

Sires all highly commended. *Romulus* won third at the London Dairy Show, in 1878, in a large ring.

Progeny at Mountainside—Heifer, June 2, 1882, Bergerelia, 2d, 19527, by Black Prince of Hanover, 2873. Sold May, 1883.

[BAR 37 MARK] *La Gallichaume, 15542.* Solid fawn, black tongue and switch; very level and square; wide-hipped, with well-sprung ribs and round barrel; head slender and delicate, with full, dark-rimmed eyes, small, incurving horns, and dark-fringed ears; straight limbs; hide mellow, loose, and rich; udder medium in size, with large teats, squarely set and well shaped.

History—Dropped spring of 1878. Bred by Wm. Amy, St. Peter's, Jersey. Sold in January, 1881, to E. P. P. Fowler, and shipped by steamer Lord Clive to Philadelphia. Bought by Mr. Havemeyer, at the Herkness May Sale the same year.

Progeny at Mountainside—Bull, January 17, 1882, *Jumbo*, 7117, by *Darlo* (F. 311), 5770. Sold June, 1882. Bull, October 17, 1882, Farmer Chaume, 9054, by *Farmer's Glory* (F. 274, H. C.), 5196. Sold May, 1883.

Royalist's Gypsy (*Pedigree Stock*), *15023*.

EAR MARK 38

Solid fawn, white tongue, black switch; a beautiful cow, very level and wide in loin; long from hips to rump, with a neat, well-poised head; up-set, widespread, fine horns; udder round, full forward, with good teats; broad escutcheon, and full milk veins; coat soft and furry; hide mellow, soft, and loose.

History—Dropped May 10, 1878. Bred by Clement Reneuf, St. Saviour's Parish, Jersey. Sold by Philip J. Mourant to E. P. P. Fowler, in January, 1881, and imported by steamship British Queen, the same month, for A. M. Herkness & Co., by whom she was sold at auction in May following.

Pedigree

FOLLOWING THE SIRE'S LINE.	COLOR.	BREEDER.	DAM.	COLOR.	DAM'S BREEDER.
Royalist's Gypsy .... 15023 .... Island name, *Gypsy*.	Solid f'n..	Clement Reneuf..	*Rosette* (F. 387 H. C.)	———..	Clement Reneuf.
Sire, *Royalist* (P. 139 H. C.)....	Gray	P. J. Mourant....	*Regina* (F. 32 H. C.).	Br'n & w.	P. J. Mourant.
2d Sire, *Duke* (P. 76 H. C.)....	Gray	Clement Lesbirel.	*Superb* (F. 353 H. C.).	Br'n & w.	C. Lesbirel.
3d Sire, *Merry Boy* (P. 61 H. C.)	Gray	John Arthur......	*Eva* (F. 628)........	Br'n & w.	John Arthur.
4th Sire, *Stockwell*, 2d (P. 24 H.C.)	Y. & w...	C. Godfrey.......	*Soucique* (F. 68)	Y. & w..	Ph. Godfrey.
5th Sire, *Noble* (F. 104 H. C.)..	Lt. br'n..	C. Pallot.........	———.............	———..	———
6th Sire, *Sultan* (F. 58 H. C.)..	Lt. br'n..	C. Pallot.........	*Flower* (F. 53 H. C.).	Br'n & w.	C. Lisbirel.
7th Sire, *Prince of Wales*......	——— ...	———...........	———.............	———..	———
8th Sire, *Old Noble*..........	——— ...	———...........	———.............	———..	———

A remarkable list of "Highly Commended" animals, both sires and dams.

ROYALIST won first prize over all Jersey in 1877; *Duke* first and sweepstakes in 1875. *Merry Boy, Stockwell,* 2d, and *Noble* were all notable prize winners.

REGINA won every possible prize on the Island "from calf to aged cow," gave 18 lbs. (Jersey weight) of butter in a week, equal to about 20 lbs. avoirdupois, and has communicated this butter-yielding quality in a remarkable degree to her descendants. Regina, 2d and 4th, are both "14-pound" cows. Dr. Howe's Cromeskin, whose record at 3 years old was 19 pounds in 7 days, is also her granddaughter.

Progeny at Mountainside—Heifer, June 1, 1881, *Rodalia*, 18443, by *Neptune* (P. 230). Sold May, 1883.
Bull, May 17, 1882, Royal Darling, 9050, by Duke of Darlington, 2460. Sold May, 1883.

EAR `[39]` **MARK**

Anna of Mountainside, 15544. Solid cream fawn, with black tongue and switch; level above and below; deep in flank, wide in loin, with broad hips and rump, a fine tail, thin neck, well-shaped, neat head, dishing face, and light horns; limbs straight and fine; a good handler, with rich skin color; udder of good form, with well-set teats.

History—Dropped in 1879. Bred by Philip Bertram, Grouville, Jersey. Sold to E. P. P. Fowler, in January, 1881, and by him imported in steamship British Queen for A. M. Herkness & Co., and sold to T. A. Havemeyer in their Sale of May following.

Pedigree—Sire, *Noble*; dam, *Beauty*.—(No further information.)

Progeny at Mountainside—Bull, May 29, 1881, *Garibaldi, 2d,* 7106, by *Garibaldi*. Sold at auction, June, 1882. Bull, April 25, 1882, Hannibal Duke, 9046, by Duke of Darlington, 2460. Sold May, 1883.

EAR 42 MARK

Gay Grisette, 9991.

Chocolate brown, with white belly, hind legs, and spot on rump; white tongue and switch. A large cow, wedge-shaped, with a straight back, high at hips; neck and shoulders thin; head narrow, but neat; horns light, black-tipped; hips and rump very wide, with long line to hocks; body deep, and limbs light and straight; udder perfect in form; teats large, square-set; quarters distinct; escutcheon first-class flanderine, wide on thighs, and running broad fully up; skin soft, mellow, and rich—the feed goes all to milk.

History—Dropped March 31, 1880. Bred by J. S. Barstow. Sold to Dr. N. J. Borland, New London, May, 1881. Bought at the Kellogg Sale in May, 1881.

Pedigree

	COLOR.	BREEDER.	SIRE.	COLOR.	SIRE'S BREEDER.
Gay Grisette, 9991........	Choc. & white..	J. S. Barstow	Lord Lee, 3104..........	Sq. gr......	J. S. Barstow.
Dam, Bonnie Grisette, 6979	Gr. f'n & w....	J. S. Barstow	Rajah of Greenvale, 2533.	D'k f'n & w.	J. S. Barstow.
2d Dam, *Grisette*, 596.....	Gray..........	F. le Sueur..	———	——— ..	———

LORD LEE'S sire, Nonquit, 1391, combines the *Primrose* and *Meg Merrilies* families, of Thos. Motley's importation, while Lady Lee, his dam, is by *RAJAH*, 340, and out of *Heather Belle*, 593.

BONNIE GRISETTE is also a granddaughter of *Rajah*, 340, through Rajah of Greenvale, out of *Fleur-de-lis*, 614.

Lady Sygny, 6938.

[43] EAR MARK

Squirrel gray, shading to mulberry fawn; white flecks on brisket and legs; white tongue and dark switch. A fine cow; level, broad, and deep, with fine tail and neat, straight limbs; a grand, well-shaped udder, with large teats and milk veins.

History—Dropped February 1, 1877. Bred by J. S. Barstow, South Portsmouth, R. I. Bought at auction, by Mr. Havemeyer, in May, 1881.

Pedigree

	COLOR.	BREEDER.	SIRE.	COLOR.	SIRE'S BREEDER.
Lady Signy, 6938	Squirrel gray	J. S. Barstow	Amir Kahn, 2573	Gray	J. S. Barstow.
Dam, Signy, 5692	Squirrel gray	J. S. Barstow	Belisario, 640	Dark gray	H. Kuhn.
2d Dam, Coral, 2707	Fawn & white	J. S. Barstow	*Rajah*, 340	Sq. gray & tan	Clement Buesnel.
3d Dam, *Clyte*, 618	—	Philip Neel	—	—	—
4th Dam, *Eliza*, 619	Lemon fawn	Philip Neel	—	—	—

Progeny at Mountainside—Bull, August 19, 1881, Babylon's King, 7109, by Babylon, 4723. Sold in 1882. Heifer, September 12, 1882, Lady Signy, 2d, 1953, by *Carlo* (P. 180 H. C.), 5559. Sold May, 1883.

EAR
| 44 |
MAINE

Hilpah, 5879. Fawn, with flecks of white on brisket and flanks, white tongue, and mixed switch. A grand cow, very level, deep, and broad, with a fine head, large, full eye, waxy horn, and dark-fringed ears; carcass capacious, broad and deep in loin and flank. She has a big udder and teats, well-shaped and well-placed, a wide selvage escutcheon, and prominent milk veins; a mellow and unctuous hide.

History—Dropped July 17, 1876. Bred by E. F. Bowditch, Framingham, Mass. Bought at the Kellogg Combination Sale, May, 1881.

Pedigree

	COLOR.	BREEDER.	SIRE.	COLOR.	SIRE'S BREEDER.
Hilpah, 5879........	Fawn & white.	E. F. Bowditch......	The Squire, 1298.	Mul. fawn..	E. F. Bowditch.
Dam, Hennie, 3335..	Sol. light fawn.	E. F. Bowditch......	Careless Boy, 1297	Dark fawn..	E. F. Bowditch.
2d Dam, Haidee, 971.	Fawn........	E. F. Bowditch......	*Sam Weller*, 271.	F'n,br'n & w.	E. F. Bowditch, impr.
3d Dam, *Hebe*, 943...	Fawn & white.	E. F. Bowditch, impr.	——	——	——

THE SQUIRE traces twice to each of those fine old cows, *Flora*, 113, and *Countess*, 114, imported by Thomas Motley, whose occurrence, combination, and recurrence in the pedigree of Jersey Belle, of Scituate, is supposed to account for her peculiar excellence. She gave 25 lbs. 3 oz. of butter in one week, and within the year 705 lbs., then unprecedented. The butter was, moreover, as yellow as gold. *Flora* gave, by Mr. Motley's test, 511 lbs. 13 oz. of butter in 50 weeks.

CARELESS BOY traces to *Flora* also, and *Sam Weller* occurs twice in the short pedigree of the cow Hennie.

Imported *HEBE* carried upon her horn the *brand* which, before Herdbook times, used to be placed upon all prize-winners on the Island.

EAR MARK 46 **Beauty of Ninon, 9691.** Silver gray, very light beneath, white tongue and switch;—level and broad, with excellent hindquarters, fine head, with small, incurving horns, long face, full, liquid eyes; hide mellow, coat furry and fine, skin-color rich, yellow in ears; good udder, and teats about the medium size.

History—Dropped December 8, 1879. Bred by Wm. Cooper, Salem, N. J. Sold to John V. N. Willis, of Marlboro, and after being a year in his herd, was put into the Kellogg Sale, and bought by Mr. Havemeyer, May, 1881.

She won *first*, as a yearling, at the Freehold, N. J., Show, in 1880.

Pedigree

	COLOR.	BREEDER.	SIRE.	COLOR.	SIRE'S BREEDER.
Beauty of Ninon, 9691..	Solid sil. gr.	Wm. Cooper......	Rory O'More, 3236...	Solid sq. gr.	East'n Exp. Farm.
Dam, Lurlei Maid, 4159.	Fawn & w..	Mrs. M. N. Rogers.	Khedive, 1051.......	Solid fawn.	S. J. Sharpless.
2d Dam, Myth, 2837....	F'n, little w.	Wm. Massey	*Count Bismarck*, 732..	Brown & w.	A. Alexandre.
3d Dam, Damsel, 2d, 1837	Squir. gray.	Wm. Massey	*St. Malo*, 486........	Sq. gr. & bn.	J. F. Page, impr.
4th Dam, *Damsel*, 1828..	Dark brown.	A. Alexandre.....	———	———	———

LURLEI MAID, as a 4-year-old, is credited with yielding 13 lbs. of butter in 6 days—equal to over 15 lbs. in a week. She traces through Khedive to *Niobe*, 99, the "Centennial Prize Cow."

Progeny at Mountainside—Heifer, June 25, 1881, Beauty of Ninon, 2d, 18444, by Pearl Rex. Sold May, 1883.
Bull, November 20, 1882, Beau-Pride, 9055, by *Farmer's Pride*, 5560. Sold May, 1883.

Jessie of Ipswich, 2883.

EAR MARK `47`

Jessie of Ipswich, 2883. Solid French gray and chocolate, black head and neck, white tongue, black switch; a long-bodied, low-set, "big-little" cow, not a beauty, but having a grand make-up for business; wedge-shaped, level-backed, high-rumped; legs short but fine and straight; udder large, fore udder extra good, milk veins prominent; skin extremely soft and mellow, well coated, and rich, but brown, not yellow; escutcheon wide on thighs, selvedge.

History—Dropped January 17, 1873. Bred by D. F. Appleton, Ipswich, Mass. From August, 1875, to October, 1880, she was the property of Mr. C. B. Morrell; returned, however, to Mr. Appleton's herd in 1880, and was entered by him in the Combination Sale, and bought by Mr. Havemeyer in May, 1881.

Pedigree

	COLOR.	BREEDER.	SIRE.		SIRE'S BREEDER.
Jessie of Ipswich, 2883........	Solid Fr. gr.	D. F. Appleton.	*Aggwam*, 597.	——..	F. P. P. Fowler, impr.
Dam, *Fanny of Ipswich*, 1650.	Mul. f'n & w.	Pierre Regnault.	——.......	——..	——

Progeny at Mountainside—Bull, February 24, 1883, Glow Boy, 9058, by *Farmer's Glory* (F. 274 H. C.), 5196. Sold May, 1883.

Dairy Maid of Bloomfield, 8352. Mulberry fawn, shading to black, with star in forehead, and white flecks on flanks; level, broad, wide, square, straight limbed; rump high; she has a neat head, with small, incurving horns, thin neck and withers; udder very capacious, with large teats; fore udder especially good; large, tortuous milk veins; escutcheon first-class flanderine; an excellent handler; skin very rich.

EAR MARK 48

History—Dropped March 22, 1878. Bred by Silas Betts, Camden, N. J. Sold to J. V. N. Willis, January, 1879; entered by him in Kellogg's Combination Sale, and bought by Mr. Havemeyer, in May, 1881.

She made as a two-year, $10\frac{1}{2}$ pounds of butter in six days, equal to $12\frac{1}{4}$ pounds in a week; she won for Mr. Willis first at Freehold, N. J., 1879 and 1880, and first at the New Jersey State Fair, in a large ring, in 1880, and was reputed a 20 quart cow.

Pedigree—

	COLOR.	BREEDER.	SIRE.	COLOR.	SIRE'S BREEDER.
Dairy Maid of B., 8352....	Fawn...	Silas Betts......	Baronet, 2240........	D'k gray.....	R. H. Stevens.
Dam, Harriet of N. J., 5949.	Gr. & w..	C. A. Cory. ...	St. Patrick, of Germantown, 2632...	D'k bronze...	J. Welch, Jr.
2d Dam, Linda 2d, 1927....	Y. fawn..	Hartman Kuhn..	Pilot Boy, 488.......	——......	H. Kuhn.
3d Dam, Linda, 846.........	——..	H. Kuhn.......	*Charleston*, 1........	F'n & w.....	H. Kuhn, impr.
4th Dam, *Princess*, 836......	D'k fawn.	H. Kuhn, impr..	——.........	——......	——

A pedigree rich with the best blood of the Pennsylvania herds, blended with choice strains from Connecticut.
ST. PATRICK, 2632, traces to *St. Helier*, 45, Taintor's *Commodore*, 56, and Norton's famous *Pansy*, 8; while Pilot Boy, 488, is a son of *Pilot*, 3, and, through his dam, traces also to *Pansy*, 8, a cow whose appearance in multitudes of the pedigrees of excellent animals is more than accidental.

Progeny at Mountainside—Heifer, January 29, 1883. Farmer's Bloom, 19535, by *Farmer's Glory* (F. 274 H. C.), 5196. Sold May, 1883.

EAR | 50 | **MARK**

Thorndale Belle, 2d, 6421.

Dark fawn, black tongue and switch; white spot in front of udder; large, wedge-shaped, deep-bodied cow, with admirable loins, great width of pelvis, hips, and rump; neat head, nearly black; hide very mellow and rich; interior of ears rich yellow.

History—Dropped April 4, 1877. Bred by Edwin Thorne. Entered in Kellogg's Combination Sale of May, 1881, and sold to T. A. Havemeyer.

Pedigree

	COLOR.	BREEDER.	SIRE.	COLOR.	SIRE'S BREEDER.
Thorndale Belle, 2d, 6421....	Solid fawn..	Edwin Thorne...	Balsora, 2357..	Solid dark gray.	B. Kittredge.
Dam, Thorndale Belle, 5265..	Very dark...	B. Kittredge.....	*Barney*, 1491..	Solid brown....	J. C. Godfrey.
2d Dam, *Kate K.*, 2d, 3726 ..	Gray & white.	Chas. de Gruchy.	——........	——........	——
3d Dam, *Kate K.*, 3717......	Gray & white.	Chas. de Gruchy.	——........	——...	——

Thorndale Belle, 5265, after milking six months, in the spring of 1881, made 14 lbs. 7 oz. of butter in 7 days.

Progeny at Mountainside—Bull, May 8, 1882, Daldimon, 9049, by Duke of Darlington, 2460. Sold May, 1883.

Mantissa, 10457. [EAR MARK: 51] Solid fawn, black tongue and switch. A grand cow, of large size and majestic bearing; very broad in loin, hips, and rump; head well shaped, with broad muzzle, strong, upset horn; barrel capacious, and well-ribbed back; udder large, well-quartered; teats medium in size, well shaped; escutcheon flanderine, well out on thighs, and running fully up, of great breadth; hide rich and unctuous.

History—Dropped December 15, 1877. Bred by Edwin Thorne. Entered in Kellogg's combination sale of May, 1881, and bought by T. A. Havemeyer.

Pedigree

	COLOR.	BREEDER.	SIRE.	COLOR.	SIRE'S BREEDER.
Mantissa, 10457	Solid light fawn.	Edwin Thorne	Cardigan, 2379	Dark gray	E. Thorne.
Dam, Kate Martin, 5267	Fawn and white.	B. Kittredge	*St. Martin*, 1482	Solid dark gray.	B. Kittredge.
2d Dam, *Kate K.*, 2d, 3726.	Gray and white	Chas. de Gruchy.	} Imported by B. Kittredge.		
3d Dam, *Kate K.*, 3717	Gray and white	Chas. de Gruchy.			

Progeny at Mountainside—Heifer, June 27, 1881, Mantissa, 2d, 18445, by Dido's Duke, 4678. Sold in 1883. Bull, February 3d, 1883, Mantrito, 9057, by *Farmer's Glory* (P. 274 H. C.), 5196. Sold May 8, 1883.

EAR
52
MARK

Lady Morgan, 11671.

Solid fawn, black tongue and switch. A well-formed cow, with a long head, up-set horns, very level and wide; deep in flank, with straight, light legs; large udder, well shaped and placed, with large teats; a wide escutcheon, extending well out upon the thighs; hide mellow and unctuous; coat soft and silky, and ear color rich and yellow.

History—Dropped June 12, 1877. Bred by Thomas Fitch, New London, Conn. Bought by Mr. Havemeyer, at the Kellogg Combination Sale, May, 1881.

Pedigree

	COLOR.	BREEDER.	DAM.	COLOR.	DAM'S BREEDER.
Lady Morgan, 11671............	Fawn & w.	Thomas Fitch...	Pierrot, 7th, 1667..	Dark gray.	Sam. C. Colt.
Dam, Princess of Mansfield, 8070.	Fawn & w.	Thomas Fitch...	Pierrot, 2d, 1669..	Fawn & w.	Sam. C. Colt.
2d Dam, Flirt, 482.............	Fawn & w.	W. W. Billings..	Santa Claus, 30...	——....	Daniel Buck.
3d Dam, Duchess, 3d, 205.......	Fawn & w.	W. W. Billings..	*Premium*, 7......	Drab & gr.	J. A. Taintor, impr.
4th Dam, *Duchess*, 2............	——....	John A. Taintor, impr.	——.......	—— ...	——

A pedigree rich in the blood of the best old Connecticut families of Jersey cattle.

Progeny at Mountainside—Bull, July 25, 1882, Morgan Carlo, 9052, by *Carlo* (P. 180 H. C.), 5559. Sold May 8, 1883.

EAR
|53| **Nameless Girl, 11623.** Chocolate fawn, white flecks on flanks and brisket, white tongue, and ring above switch. A beautiful cow, with neat head and horns; level and deep, with capacious barrel; loin and hips of medium width, rump slightly depressed, but tail set level; hide mellow, flexible and loose; coat soft, and skin color rich; udder of medium size, well-shaped, with good teats and a fair escutcheon.
MARK

History—Dropped March 31, 1879. Bred by E. F. Bowditch, Framingham, Mass. Bought by Mr Havemeyer at the Kellogg Combination Sale of May, 1881.

Pedigree

	COLOR.	BREEDER.	SIRE.	COLOR.	BREEDER.
Nameless Girl, 11623.....	Chocolate..	E. F. Bowditch..	Fast Boy, 2606...	Gray fawn & w..	C. Wellington.
Dam, Nimmie of M., 4433.	Fawn & w.	E. F. Bowditch..	The Squire, 1298.	Solid mul. fawn.	E. F. Bowditch.
2d Dam, Nimmie, 968....	———.....	E. F. Bowditch..	*Sam Weller*, 271.	Fawn and white.	E. F. Bowditch, impr.
3d Dam, *Nora*, 956......	Fawn & w.	E. F. Bowditch, impr.	———	———	———

NIMMIE OF MILLWOOD traces twice to Motley's *Flora*, 113, twice to *Countess*, 114, and to other strains of blood rendered famous of late by their combination and recurrence in the pedigree of the noted cow JERSEY BELLE OF SCITUATE, 7828, whose yield of 25 lbs. 3 oz. of butter in 7 days and 705 lbs. in one year was for a time unequalled. *Flora*, 113, yielded, by Mr. Motley's test, 511 lbs. 1 oz. of butter in 50 weeks.

Progeny at Mountainside—Bull, March 15, 1882, Bow-Carlo, 9042, by *Carlo* (P. 180 H. C.), 5559. Sold May 8, 1883.

EAR MARK [55]

Flora D., 11192.
Solid, dark fawn, shading to gray; white tongue, black switch. A grand, deep-bodied, broad, level, fine-limbed cow, with excellent hindquarters, good head, and thin neck and withers; a good handler; udder well shaped, teats medium, milk veins prominent, escutcheon fair.

History—Dropped August 20, 1879. Bred by C. Deming, Farmington, Conn.; transferred to McKeen & Hulick, December, 1880; bought at auction by T. A. Havemeyer, May, 1881.

Pedigree

	COLOR.	BREEDER.	SIRE.	COLOR.	SIRE'S BREEDER.
Flora D., 11192	Solid fawn . .	C. Deming	Tunxis Chief, 3705 .	Sm'ky f'n.	C. Deming.
Dam, Deming's Flora, 4398	Fawn & w . .	H. W. Bassett . . .	Ned Booth, 1508 . . .	Br. gr. & w	P. B. Tyler.
2d Dam, Dickinson's Belle, 4395 .	D'k f'n & w .	S. P. Dickinson . .	McClellan, 3d, 27 . .	Fr. gray . . .	J. T. Norton.
3d Dam, E. Colt's Belle, 4167 . . .	Fawn	Elisha Colt	Bill, Jr., 182	D'k brown.	Samuel Colt.
4th Dam, *Violet*, 23	——	John A. Taintor, impr.	———	———	——

A first-rate old-fashioned Connecticut pedigree, running back eight or ten generations on certain lines and including, often repeated, such grand animals as *Pansy*, 8, *Splendid*, 2, Angelina Baker, *Bill*, 50, *Jenny*, 16, etc., of the Norton and Taintor and Colt importations, than which no better were ever made.

Progeny at Mountainside—Flora D., 2d, 18448, Heifer, November 1, 1881, by *Lord Clive*, 3313. Sold in 1883.

Flora D., 3d, 19532, Heifer, October 22, 1882, by Black Prince of Hanover. Sold May 8, 1883.

EAR MARK 56 Token, 10262. Solid gray fawn, black tongue and switch. A cow of large size, level and even throughout; very deep in flank, with good head, limbs, and tail, but strong framed all over; udder large, soft, slightly coupé, teats of medium size, escutcheon flanderine, skin mellow, and of good color.

History—Dropped June 30, 1879. Bred by Thomas J. Hand. Sold to Mr. Burnham, of Saugatuck, Conn., by him entered in the Kellogg Combination Sale, May, 1881, and bought by Mr. Havemeyer.

Pedigree

	COLOR.	BREEDER.	SIRE.	COLOR.	SIRE'S BREEDER.
Token, 10262	Gray fawn	T. J. Hand	Butter Boy, 3243	Solid fawn	John D. Wing.
Dam, Symbol, 6136	Solid gray	T. J. Hand	Lord Lawrence, 1414	Dark Fr. gray.	T. J. Hand.
2d Dam, Empresa, 2790	Solid	T. J. Hand	Marius, 760	Solid gray	T. J. Hand.
3d Dam, *Emblem*, 90	Gray & white	E. Gibant	*Clement*, 115	Fawn & white.	J. H. McHenry, impr.

TOKEN'S pedigree combines blood of rare excellence, especially that of LADY MARY, 1148 (the dam of Marius and of Lord Lawrence—the former by her own son, Willie Boy, and himself, Marius, the sire of SIGNAL, the famous butter bull), of *Emblem*, 90, and of *Lawrence*, 61, the foundation animals of Mr. Hand's herd. She gets good blood also on her sire's side, Butter Boy, tracing on the dam's line to Faile's *Edith*, and Hoe's *Saturn*, and on his sire's line to Kittredge's *Beauty*, 534.

Progeny at Mountainside—Bull, November 12, 1881. Token's Ossian, 8503, by Ossian, 3249. Sold May, 1883.

EAR 57 MARK

Phyllis, 10657. Solid dark gray, black tongue and switch. A rather tall, medium-sized cow, with a neat head, thin, clean neck and chine; back straight, with broad loin and pelvis, a close twist, and thick thighs; hide mellow, coat silky, with rich skin color; fair udder; fore udder and teats especially good and large.

History—Dropped June 9, 1880. Bred by J. V. N. Willis, Marlboro, N. J. Won *first* at the New Jersey State Show in 1880. Entered in the Kellogg Combination Sale, and bought by Mr. Havemeyer, May, 1881.

Pedigree

	COLOR.	BREEDER.	SIRE.	COLOR.	SIRE'S BREEDER.
Phyllis, 10657......................	Solid f'n.	J. V. N. Willis..	Castle Boy, 3517..	Solid f'n.	Mrs. M. N. Rogers.
Dam, Dairy Maid of Bloomfield, 8352.	Fawn ...	Silas Betts......	Baronet, 2240.....	D'k gray.	R. H. Stevens.
2d Dam, Harriet of N. J., 5949......	Gr. & w.	C. A. Cory......	St. Patrick of Germantown, 2632..	D'k br'ze	J. Welch, Jr.
3d Dam, Linda, 2d, 1927	Yel. f'n..	Hartman Kuhn..	Pilot Boy, 488	———..	H. Kuhn.
4th Dam, Linda, 846...............	———..	Hartman Kuhn..	*Charleston*, 1......	F'n & w.	H. Kuhn, impr.
5th Dam, *Princess*, 836	Dark f'n.	Hartman Kuhn, impr.			

A number of excellent animals give tone to this pedigree—*Niobe*, 99, the Centennial prize cow; *Europa*, 121; *Lilly*, 1; McClellan, 25; Albert, 44; *St. Helier*, 45—aside from those mentioned in the direct line of dams and their sires.

EAR [58] **MARK**

Favorite's May, 8662. Solid cream fawn, white tongue, and dark switch; very level, deep in the flanks, but round in the barrel; thick through the heart. A fine large cow, with a noble head; a good handler, with a furry coat, soft and rich, with yellow skin color; limbs neat and straight; udder of grand size—fore udder very fine.

History—Dropped July 6, 1879.   Bred by John I. Holly, Plainfield, N. J.; sold at Kellogg's Combination Sale, May, 1881.

Pedigree—

	COLOR.	BREEDER.	SIRE.	COLOR.	SIRE'S BREEDER.
Favorite's May, 8662	Solid f'n	John I. Holly	Much Ado, 2405	Solid f'n	J. B. Williams.
Dam, *Favorite of the Elms*, 1656.	Fawn	Peter le Gresley	——	——	——.

FAVORITE OF THE ELMS is a cow of extraordinary perfection of form and usefulness. She has been often exhibited, and, we believe, never beaten in the show ring, having won the New Jersey State special first in 1874, the New Jersey State Society's first in 1876, was in the first-prize herd in 1877, 1878, and 1879, and won the sweepstakes in 1879, all at Waverly. Besides, she won three first prizes at the Mt. Holly Show, and first at the Dairy Show held at the American Institute Building in 1879.

MUCH ADO is by Dash of Glastenburg, 1959, out of a 26-quart cow, *Dandelion*, 2521, whose 4-year-old daughter is said to have made 14 lbs. of butter in 7 days.   Dash of G. traces to Albert, 44, and Splendid, 2, on the sire's side, and in the line of his dam's sire, Tom Dasher, to Albert again, McClellan, 25, and Pansy, 8.   Good enough.

Progeny at Mountainside—Heifer, July 17, 1882, Favorite's May, 2d, 19529, by Gilderoy, 2107.   Sold at auction, May, 1883.

EAR 59 MARK

Malita, 5169. Dark fawn and white, large speckled spots on thighs, fleck at setting on of tail, white tongue, and mixed switch. A medium-sized, well-formed cow. She has a handsome, dark head, shading to gray, a long face and fringed ears, and a neat, waxy horn; a straight and deep carcass, broad pelvis, well-shaped udder and teats, and good skin color.

History—Dropped August 24, 1876. Bred by J. Carter Brown, East Greenwich, R. I. Sold at the Combination Sale, May, 1881, and bought by T. A. Havemeyer.

Pedigree

	COLOR.	BREEDER.	SIRE.	COLOR.	SIRE'S BREEDER.
Malita, 5169	D'k f'n & w.	J. Carter Brown........	Alpheus, 1168....	D'k f'n.	R. M. Hoe.
Dam, *Miss McMahon*, 2739......	——....	Capt. F. Perrée........	*Gen. McMahon* (P. 32).	Solid..	E. J. Simon.
2d Dam, *Madeleine, 2d*, 2479.....	Solid......	J. Carter Brown, 2d impr.	*Pierrot* (F. 143)	Lt. gr.	Jas Ballaine.
3d Dam, *Madeleine* (F. 651 H. C.)	Br.,gr.,& w.	W. Luckarift.			

ALPHEUS is full brother to Sarpedon—pure "Alphea," with no out-cross. The rest of the pedigree is Jersey bred, and very fine. *Pierrot* was subsequently imported by Sam. C. Colt, and for years led his fine herd with distinction.

Satin, 10329. [EAR MARK: 60] Solid chocolate fawn, black tongue and switch; very level, compact, and deep; small-boned, and well knit every way; barrel with well-sprung ribs, and unusually deep and capacious; loin and hips broad; long to rump, which is wide, with the tail set level and square; a grand cow every way; udder enormous, with large teats; large, full, and tortuous milk-veins; escutcheon broad, and ascending fully up, double-selvedge class; hide flexible and mellow, velvety, and very rich in color and unctuousness.

History—Dropped November 26, 1878. Bred by A. B. Darling. Bought at the Kellogg Combination Sale, May, 1881.

Pedigree

	COLOR.	BREEDER.	SIRE.	COLOR.	SIRE'S BREEDER.
Satin, 10329......	Solid choco. f'n.	A. B. Darling...	Duke of Darlington, 2460.	Gray......	A. B. Darling.
Dam, Oriole, 2563..	Dark fawn & w.	R. M. Hoe......	*Dolphin*, 2d, 468........	Gray......	F. M. Wilson, Eng.
2d Dam, Leda, 799..	Solid dark fawn.	R. M. Hoe......	Jupiter, 93.............	Brown & w.	R. M. Hoe.
3d Dam, Europa, 176	Brown fawn....	R. M. Hoe......	Jupiter, 93.............	Brown & w.	R. M. Hoe.
4th Dam, Alphea, 171	Brown fawn....	R. M. Hoe......	*Saturn*, 94.............	Reddish f'n	R. M. Hoe, impr.
5th Dam, *Rhea*, 166.	Dark brown....	R. M. Hoe, impr.	———	———	———

DUKE OF DARLINGTON, 2460, is the most famous son of Eurotas, 2454, perhaps the most famous living Jersey. He is also the sire of Bomba, 10330.

RIOTER, 2d, 467, in the pedigree of Eurotas, and *Dolphin*, 2d, 468, the sire of Oriole, 2563—two English-bred bulls of the Dauncey blood—are the only out-crosses (and valuable ones they are) in this otherwise "pure Alphea" pedigree.

Progeny at Mountainside—Bull, Silver Sheen, 9047, May 4, 1882, by Black Prince of Hanover, 2873. Retained in the herd—No. 60—1, page 13.

Amulet of Home Farm, 12096.

EAR MARK 61

Yellow fawn, with white star in forehead, spot on left flank, white tongue and switch. A comparatively small cow, strongly wedge shaped, straight to hips, drooping to the rump. She has a pretty head, with fringed ears, upset horns, light and rich, a clean, thin neck, broad, straight back, and large hindquarters; legs and tail fine; a mellow hide, silky coat, and rich skin color; selvedge escutcheon, and well-shaped udder and teats.

History—Dropped October 8, 1880. Bred by Thos. J. Hand, Sing Sing, N. Y.; sold in her dam to H. S. Russell, Milton, Mass., and sold by him at Kellogg's Combination Sale of May, 1881.

Pedigree

	COLOR.	BREEDER.	SIRE.	COLOR.	SIRE'S BREEDER.
Amulet of H. F., 12096....	Yel. f'n.	T. J. Hand........ ..	King Horn, 3280......	Solid........	T. J. Hand.
2d Dam, Witch Hazel, 1360.	F'n & w.	T. J. Hand..........	*Southampton*, 117.... .	Or. br'n & bl'k	Philip Gaudin.
3d Dam, *Hazel*, 91........	F'n & gr.	J. H. McHenry, impr.	*Clement* (F. 61 H. C.)..	D'k f'n & w..	Ed. Gibaut.
4th Dam, *Lady Bird*.......	Gr. & w.	J. H. McHenry, impr.	*King Hal* (F. 92)......	Dark brown..	H. J. le Feuvre.

KING HORN is a son of Hornbeam, by Marius, both son and grandson of *Lady Mary*, his (King Horn's) dam. Hornbeam's dam, Emily Hampton, duplicates the blood of *Southampton* and introduces that of *Emblem*, both animals of fame.

EAR MARK | 62 |

Brownie, 1166. Dun and white, white spot on forehead, shoulder stripe, white tongue and switch. A grand old cow, well formed, wedge-shaped, broad, deep, capacious, strong-constitutioned; an ideal "milk-producing machine;" udder enormous, with large teats; milk veins large and tortuous; escutcheon, broad curveline; hide mellow, and skin color rich orange.

History—Dropped May 11, 1868. Bred by Benj. L. Swan, Jr., Oyster Bay, N. Y. Bought at the Kellogg Combination Sale, May, 1881.

Pedigree

	COLOR.	BREEDER.	SIRE.	COLOR.	BREEDERS AND IMPORTERS.
Brownie, 1166	Dun & w	Benj. L. Swan, Jr.	*Express*, 328	Dark fawn	John Hoey, impr.
Dam, Edith, 871	Orange f'n	Benj. L. Swan, Jr.	*Black Roger*, 326	Tan & black	Wm. Redmond, impr.
2d Dam, Gazelle, 873	Lt. f'n & w	Wm. B. Bacon	Major, 3d, 230	——	S. Henshaw.
3d Dam, *Buttercup*, 872	——	Wm. B. Bacon, Impr.	——	——	——

Progeny at Mountainside—Bull, Brown Carlo, June 14, 1882, by *Carlo* (P. 180 C.), 5559. Sold May, 1883.

Nellie Rival, 11147. Solid dark fawn; full black points, tongue, and switch. A beautiful cow, of medium size, very neat and deer-like, level, broad loined and hipped, with fine limbs; udder well shaped and capacious, with medium-sized teats.

EAR MARK: 64

History—Dropped August 19th, 1880, on farm of F. C. Havemeyer. Bred by Wm. Crozier. Sold in her dam to F. C. Havemeyer; transferred by him to T. A. Havemeyer in May, 1881.

Pedigree

	COLOR.	BREEDER.	SIRE.	COLOR.	BREEDER.
Nellie Rival, 11147	Solid fawn..	Wm. Crozier..	*Rival* (P. 143), 3762..	Sol. yel. f'n..	W. Alexandre.
Dam, *Nellie le Brocq*, 8986.	Bronze	F. le Brocq...	———	———......	———

Progeny at Mountainside—Bull, Unrivalled, May 7, 1882, by *Farmer's Pride*, 5560. Sold May, 1883.

EAR 65 MARK **Bay Queen, 11145.** Solid dark fawn, gray head, black tongue, mixed switch. A medium-sized cow, straight, slender, and stylish, with a fine head, neat, incurving, up-set horn, thin neck and withers, deep flank, excellent hindquarters, fine tail, and deer-like legs; hide loose and mellow, with silky coat and rich, creamy skin color; udder and teats of excellent form and good size.

History—Dropped April 29, 1880. Bred by F. C. Havemeyer, Throgg's Neck, N. Y., and transferred to T. A. Havemeyer in May, 1881.

Pedigree

	COLOR.	BREEDER.	SIRE.	COLOR.	SIRE'S BREEDER.
Bay Queen, 11145	D'k fawn	F. C. Havemeyer.	Nobleman of the Elms.	Solid gray	W. S. Taylor.
Dam, *Miss Browney*, 7288	Solid	Francis le Maistre.	*Browney* (P. 158)	Solid lt. br'n.	Wm. Avril.
2d Dam, *Miss De Cartaret*, 7285.	Bronze, gr. & w.	Francis le Maistre.	*Nelson* (P. 65)	Y. f'n & w.	P. du Val.
3d Dam, *Florence*	—	—	—	—	—

NOBLEMAN OF THE ELMS is a son of that remarkably beautiful and useful "16-lb." cow, *Favorite of the Elms*, long the queen of William S. Taylor's herd, and now gracing the herd of John I. Holly.

Lady Hammond, 7286.

EAR MARK | 66

Solid fawn, black points. A noble cow, very broad and deep, with a beautiful head and horns, excellent limbs, strong and straight; deep in brisket and in flank; grand hindquarters, large udder and teats, broad flanderine escutcheon. An excellent handler, with unctuous and yellow skin.

History—Dropped May 13, 1878. Bred by F. C. Havemeyer, and transferred to Mountainside Herd, May, 1881.

Pedigree

FOLLOWING THE LINE OF SIRES.	COLOR.	BREEDER.	DAM.	COLOR.	DAM'S BREEDER.
Lady Hammond, 7286........	Solid f'n.	F. C. Havemeyer.	*Lily Cabot*, 6325	Lem. f'n.	J. Cabot.
Sire, Sylvester, 2d, 2851.......	Gray f'n.	F. C. Havemeyer.	*Canary of Jersey*, 6324.........	Solid f'n.	P. de la Parelle.
2d Sire, Sylvester, 694........	Solid d'k.	Herbert Mead....	*Sylphide*, 169..................	Red & w.	R. M. Hoe, imp.
3d Sire { *Dolphin, 2d, 468..* *Young Dolphin* (247 E. H. B.)............ }	F'n, sil. gr.	F. M. Wilson, Eng.	*Vanity*, by *Pedlar* (631 E. H. B.).	——....	P. Dauncy.
4th Sire, *Dolphin* (242 E. H. B.)	Gray....	P. Dauncy.......	*Doll*, by *Vapour* (906 E. H. B.)....	——....	P. Dauncy.
5th Sire, *Wapiti* (927 E. H. B.).	Solid....	P. Dauncy.......	*Wapiti*, by *Bodachglas* (927 E. H. B.)——......		P. Dauncy.
6th Sire, *Pedlar* (631 E. H. B.).	Solid....	P. Dauncy	*Label*, by *Fortune* (332 E. H. B.)...	——....	———
7th Sire, *Paladin* (618 E. H. B.).	Solid....	P. Dauncy.......	*Paragon*, by *Fortune* (332 E. H. B.)	——....	———
8th Sire, *Vapour* (906 E. H. B.)	Solid....	P. Dauncy.......	*Violet*....................	Solid col.	Col. le Couteur.

This pedigree is noticeable from the fact that though bred for several generations in this country, while the sires were American-bred, the dams were all imported.

DOLPHIN, 2d, 468, is one of two English-bred bulls imported by Colonel Hoe, and which made a great impression for good upon his herd. He springs from the Dauncy Herd, which was the most famous among English Jersey herds.

Progeny at Mountainside

—Heifer, November 15, 1880, Lola Hammond, 12252, by Lord Starr, 3746. No. 86

Heifer, March 3, 1872, Lady Hammond, 2d, 18457, by *Prince Hammond*, 3672. Sold at auction, May, 1883.

EAR
|67|
MARK

Ochra, 2845. Light fawn, very even and rich, head gray, white switch, and patch on brisket. A large, noble cow, very level, deep bodied, thick through the heart, with a long face, small, black, up-set horns, and thin neck; excellent udder and large teats; hide mellow, with rich skin color.

History—Dropped August 17, 1875. Bred by C S Sargent, Brookline, Mass. Sold to Wm. Crozier, May 2, 1877; to F. C. Havemeyer, Dec. 19, 1879; and to T. A. Havemeyer, June 13, 1881.

Pedigree

	COLOR.	BREEDER.	SIRE.	COLOR.	SIRE'S BREEDER.
Ochra, 2845	Fawn	C. S. Sargent	Nonquit, 1391	Solid gray fawn	C. S. Sargent.
Dam, Azalia, 1443	Solid	C. S. Sargent	*John le Bas*, 398	Brown and white	Thos. Motley, impr.
2d Dam, *Nellie*, 289	Mouse color.	Thos. Motley, impr.			

EAR MARK 68

Alcmena, 6193.

Solid squirrel gray, white tongue, black switch. A fine large cow of grand constitution, very straight and broad, wide-hipped, wedge-shaped, deep in the flank, fine in bone, and handsome from every point of view ; udder large, milk veins tortuous and prominent ; hide exceedingly soft and mellow, and the color in ears and of the skin a rich orange.

History—Dropped September 5, 1877. Bred by William S. Taylor, Burlington, N. J. Sold to F. C. Havemeyer June 22, 1878, and transferred to the Mountainside Herd January 13, 1881.

Pedigree

	COLOR.	BREEDER.	SIRE.	COLOR.	SIRE'S BREEDER.
Alcmena, 6193............	Sol. gr. f'n.	Wm. S. Taylor..	*Nobleman*, 2d (P. 123), 2409	Sol. gr. f'n	John le Brun.
Dam, Corsetta, 1859.......	Y. f'n & w.	W. C. Wilson...	Viceroy, 346................	——...	S. J. Sharpless.
2d Dam, Leda, 149	Lt. f'n & w.	W. C. Wilson...	Prince, 55	Lt.f'n & w.	W. C. Wilson.
3d Dam, *Clara*, 148.......	D'k f'n & w.	John A. Taintor, impr.	——...	——...	——

Progeny at Mountainside—Heifer, December 15, 1882. Alcmena, 4th, 19533, by *Farmer's Glory* (P. 274 H. C.), 5196. Sold May 8, 1883.

EAR MARK [69] *Brunette Hammond,* 7284. Solid fawn, shading to brown, black tongue and switch; large, deep, level, thin necked, broad loined, high rumped, and fine boned; face wide between the eyes, with dark shadings; fringed ears, and even, incurving, drooping horns. A beauty, with a superb udder, very full forward, full milk-veins, and a fine, thin, mellow skin.

History—Dropped February, 1876. Bred by Philip Pinet, St. John's, Jersey. Imported in steamship Otranto in March, 1878, by E. P. P. Fowler, and bought at auction the same month by F. C. Havemeyer. Transferred to T. A. Havemeyer June 13, 1881.

Pedigree—*Brunette Hammond*, 7284. Fawn. E. P. P. Fowler, impr. Sire, *Milord*. Dam, *May Flower*.

Progeny at Mountainside—Bull, January 13, 1882. Brunette's Prince, 7115, by *Prince Hammond*, 3672. Sold May, 1882.

EAR MARK 70

Annie Gold-Dust, 6849.

Solid dark fawn, inclining to gray ; black tongue and switch. Not a large cow, but extraordinary for depth and width of carcass, depth of flank, and substance. She is level and square, fine-boned, with a neat head and horn. She has a large, well-shaped udder, an irregular escutcheon, which is both broad and high. Her hide is mellow and loose, coat silky and soft.

History—Dropped January 16, 1878. Bred by H. S. Parke, Bayside, L. I. Bought at auction, October 13, 1880, by F. C. Havemeyer, and transferred to Mountainside Farm, June 13, 1881.

Pedigree

	COLOR.	BREEDER.	SIRE.	COLOR.	SIRE'S BREEDER.
Annie Gold Dust, 6849....	Solid dark f'n.	H. S. Parke.......	*Jersey Gold Dust*, 2134.	Cream....	Wm. Amy.
Dam, Annie Page, 2690..	Solid fawn....	H. S. Parke.......	Son of Alphea, 562....	Solid fawn.	R. M. Hoe.
2d Dam, *Princess*, 1154...	———.......	W. H. T. Hughes, impr.	———............	———....	———

Her dam, ANNIE PAGE, was discovered to have an udder full of milk at eleven months old, ten months and more before she calved, and was regularly milked with the other cows after that. She proved a deep and rich milker.

SON OF ALPHEA was the last son of that wonderful cow Alphea, 171, whose yield of butter is estimated at 25 pounds a week, with strong evidence that it is not an over-estimate.

Progeny at Mountainside—Heifer, March 20, 1881, Fairy Gold-Dust, 14612, by *Jersey Gold-Dust*, 2134. Heifer, February 14, 1882, Annie Gold-Dust, 2d, 18455, by *Prince Hammond*, 3672. Sold May, 1883. Heifer, December 21, 1882, Annie Gold-Dust, 3d, 19534, by *Farmer's Glory* (P. 274 H. C.), 5196. Sold at auction in May, 1883.

EAR
71
MARK

Lille Bonne, 8108. Solid light brown, with dark-fringed ears, black tongue and switch. One of the noblest cows in the herd; large, and very handsome; head long, narrow, dished, with full, mild eyes, small, waxy horns; back level, loin and hips broad, rump high, body very deep and capacious; limbs fine and straight; udder very large, with large, well-placed teats; skin soft, mellow, and rich in color.

History—Dropped August 12, 1876. Bred by Francis le Brocq, St. Peter's, Jersey. Sold to E. P. P. Fowler, and imported by him, September, 1878. Bought at auction by F. C. Havemeyer, October, 1878. Sold to Wm. Crozier, November, 1879. Bought back in January, 1880, and transferred to Mountainside Herd, June 13, 1881.

Pedigree

	COLOR.	BREEDER.	SIRE.	COLOR.	SIRE'S BREEDER.
Lille Bonne, 8108....	Solid brown......	Fr. le Brocq...	*Plough Boy* (P. 102)..	Light fawn..	Wm. Luckarift.
Dam, *Queen* (F. 1239)	Yellow and white.	W. Kerslake..	——————............	——————......	——————

Progeny at Mountainside

Heifer, November 15, 1880, Lille Bonne, 2d, 12253, by Lord Starr, 3746. No. 87 of the herd.
Heifer, September 3d, 1881, Lille Bonne, 3d, 18446, by *Prince Hammond*, 3672. Sold May, 1883.
Bull, March 3, 1883, Lille Bonne's Glory, 9115, by *Farmer's Glory*, 5196. Sold May, 1883.

EAR
72
MARK

Fleur de Leury, 19651. Solid brown, black tongue and switch. A cow of rare beauty, with an elegant stylish head, neat horns, thin neck and withers, level and capacious body, well-ribbed back, broad loin and hips, long and wide rump, long thighs, and deep flanks ; brisket and belly low and level ; udder and teats large—fore-udder especially large and well formed ; hide mellow and skin rich and unctuous.

History—Dropped February 5, 1880.   Bred by Philip Labey, Grouville, Jersey.

Won, as *Fleur de Lis, 2d, first* at Grouville Parochial Show, 1881, and *first* over all Jersey in yearling class the same year.   Bought by Mr. Burnett, for T. A. Havemeyer, in August, 1881, and imported in steamship Holland the same month.

Pedigree

FOLLOWING THE LINE OF SIRES.	COLOR.	BREEDER.	DAM.	COLOR.	DAM'S BREEDER.
Fleur de Leury, Island name, *Fleur de Lis, 2d,* } 19651..........	Brown....	Philip Labey....	*Fleur de Lis* (F. 1963) ..	——.	Philip Labey.
Sire, *Farmer's Glory* (F. 274 H. C.)..	Gray......	F. le Brocq.....	*Bonheur* (F. 1651 H. C.)	Lt. br'n..	F. Becquet.
2d Sire, *Grey King* (P. 169 II. C.)..	Silver gray.	Wm. Alexandre.	*Lily Grey* (F. 770)	Silver gr..	Wm. Alexandre.
3d Sire, { *Duke* (P. 76 H. C.)... *Sweepstakes Duke,* 1905 }	Gray......	Clement Lesbirel.	*Superb* (F. 353 H. C.)...	Br'n & w.	C. Lesbirel.
4th Sire, *Merry Boy* (P. 61 H. C.) .	Gray......	John Arthur. ..	*Eva* (F. 628)...........	Br'n & w.	J. Arthur.
5th Sire, *Stockwell* (P. 24 H. C.)....	Y. & w....	C. Godfrey.....	*Soucique* (F. 68)........	Yel. & w.	Ph. Godfrey.
6th Sire, *Noble* (F. 104 H. C.).....	Lt. brown.	C. Pallot.......	——..............	— ...	——
7th Sire, *Sultan* (F. 58 H. C.)......	Lt. brown.	C. Pallot.......	*Flower* (F. 53 H. C.).....	Br'n & w.	C. Lesbirel.
8th Sire, *Prince of Wales*..........	——....	....	——...............	—— ...	——

FLEUR DE LIS (F. 1963), being of the Foundation Stock of the Island, has no recorded pedigree.

FARMER'S GLORY bequeathes beauty and honor in the show ring, with abundance and richness at the pail.

EAR MARK 73

Jessy Belle, 19647. Brown fawn, dark head, white on brisket, white tongue, mixed switch. A grand cow, of large size, broad, level, deep, and fine. She has a short head, broad and dished, with full, placid eyes, light horns, and rich, dark-fringed ears; back straight, highest at rump; carcass cylindrical, well-ribbed back, deep in flank, with low brisket and nether line; hindquarters well developed; udder excellent, exceptionally so forward, with large teats; milk veins prominent; selvedge escutcheon; hide mellow and loose, with close coat and rich yellow skin color.

History—Dropped January 20, 1879. Bred by Ph. le Feuvre, St. Ouen's Parish, Jersey. Bought by E. Burnett for T. A. Havemeyer, and imported in steamship Holland to New York, in August, 1881.

She won the *third* prize, as a yearling, over all Jersey in May, 1880, and *second* prize over the Island as a two-year-old, in May, 1881.

Pedigree

	COLOR.	BREEDER.	SIRE.	COLOR.	SIRE'S BREEDER.
Jessy Bell, 19647........	Brown..	Ph. le Feuvre..	*Browny* (P. 158)....	Solid light brown....	William Avril.
Dam, *Timbuctoo* (F. 2525)	——..	Ph. le Feuvre..	——............	——............	——

EAR
74
MARK

Victor's Lucie, 19649. Solid dark fawn, black tongue and switch, very broad and level; deep carcass, wide-hipped, slightly depressed rump, strong, straight limbs, neat head and horn; an excellent handler; skin mellow, unctuous, and exhibiting moderate richness of color; udder and teats well shaped; escutcheon broad.

History—Dropped March 12, 1879. Bred by Philip le Brocq, St. Mary's, Jersey. Bought by E. Burnett, for T. A. Havemeyer, and imported to New York in steamship Holland, in August, 1880.

Pedigree

	COLOR.	BREEDER.	SIRE.	COLOR.	SIRE'S BREEDER.
Victor's Lucie, 19649.	Solid dark fawn..	Ph. le Brocq....	*Victor* (P. 148 H. C.)...	Brown..	F. le Brocq, Jr.
Dam, *Lucie* (F. 1385)..	——	——	——	—— ..	——

Progeny at Mountainside—Heifer, *Venus Victrix*, dropped at sea. Sold at auction, May, 1883.

[EAR 76 MARK] *Miss Alexandre, 00000.* Solid brown, shading to gray, black tongue and switch. A fine, large-sized, useful cow, with handsome head, thin neck and withers; back level, with loin and hips of good width; frame well knit, and indicating good constitution; legs straight and fine; hide mellow, soft, and rich, with fine yellow color showing in ears and upon the udder, which is large and well formed; milk veins prominent and escutcheon broad, horizontal with thigh ovals.

History—Dropped February, 1879. Bought of Wm. Alexandre, St. Martin's Parish, and probably bred by him. Selected by Edward Burnett, for T. A. Havemeyer, and imported in steamship Holland to New York in August, 1881.

Pedigree

	COLOR.	BREEDER.	SIRE.	COLOR.	SIRE'S BREEDER.
Miss Alexandre, 00000	Solid brown	Wm. Alexandre	*Beauty*	———	Miss Gaudin.
Dam, *Jessie*	———	John le Huguet	———	———	———

Miss Huelin, 22996.

EAR
| 77 |
MARK

Miss Huelin, 22996. Dark fawn, fringed ears, black tongue and switch. A large cow—level, broad, and wide hipped ; rump slightly depressed ; head neat, with black horns, muzzle broad, face long ; barrel capacious, with well-sprung ribs and deep flanks ; udder well shaped, large, with large teats ; escutcheon well out on thighs, high and broad ; skin mellow and rich.

History—Dropped April, 1879. Bred by John Huelin, St. Mary's Parish, Jersey. Sold to E. Burnett, for T. A. Havemeyer, August, 1881. Imported in steamship Holland to New York the same month.

Pedigree—

	COLOR.	BREEDER.	SIRE.	COLOR.	SIRE'S BREEDER.
Miss Huelin, 22996......	Dark fawn....	John Huelin....	Victor (P. 148 H. C.)..	Brown..	F. le Brocq, Jr.
Dam, *Mignonne du Clos Failu,* (F. 4726)....	——.......	John Huelin....	——.	——..	——

Progeny at Mountainside—Bull, December 2, 1882, Mountain Glory, oooo, by *Farmer's Glory* (P. 274 H. C.), 5196. Sold May, 1883.

EAR MARK 78 *Landseer's Dove (P. 415), 19648.* Nearly solid gray fawn; small white patch on right arm, and under fore leg on brisket; black tongue, mixed switch; a large, straight-bodied, well-formed cow, with broad, deep, and capacious body; good, straight limbs, a strong horn, but neat and waxy-yellow; an udder of good capacity, slightly *coupé* forward, with large, well-shaped teats; a good handler—skin very unctuous and rich in color; escutcheon broad curveline.

History—Dropped February 17, 1879. Bred by Philip le Feuvre, St. Ouen's, Jersey. Qualified for admission to the Herdbook, July 14, 1881. Bought by E. Burnett for T. A. Havemeyer, in August, 1881, and imported by ship Holland to New York the same month.

Pedigree

		COLOR.	BREEDER.	SIRE.	COLOR.	SIRE'S BREEDER.
Landseer's Dove, Island name, "*Dove.*"	(P. 415)	Gray fawn	Ph. le Feuvre	*Landseer* (P. 162 C.)	Gray	Wm. Luckarift.
Dam, *Blooming* (F. 921 H. C.)		Yel. fawn	P. J. Mourant	——	——	——

Progeny at Mountainside—Bull, June 3, 1882, Bloom Noble, 9155, by *Prize Noble*, 7321. Sold at the auction sale of May, 1883.

EAR
79
MARK

Art Souvenir, 22997. Solid fawn, black tongue and switch; head gray. A very superior cow, with straight back and wide loin; hindquarters exceedingly well developed, hips wide, rump long and wide, thighs long, tail slender, bone fine throughout; legs straight and fine, belly and flank level and moderately deep; hide thin and loose, coat furry, skin color in ears, etc., rich yellow; udder of excellent, even form, with good-sized teats; escutcheon broad curveline.

History—Dropped January, 1880. Bred by Francis le Feuvre, St. Ouen's, Jersey. Bought by E. Burnett, for T. A. Havemeyer, in August, 1881, and imported to New York by steamship Holland the same month.

Pedigree

	COLOR.	BREEDER.	SIRE.	COLOR.	SIRE'S BREEDER.
Art Souvenir; 22997. Island name, *Souvenir, 2d.*	Solid fawn...	Francis le Feuvre..	*Artist* (F. 276 C.)......	Gray.....	Wm. Amy.
Dam, *Souvenir.* (F. 2061).	———......	Francis le Feuvre..	———...........	——— ...	———

ARTIST won *first* at St. Peter's Parochial Show in 1880.

Progeny at Mountainside—Bull, October 12, 1882, Gift, 0000, by *Farmer's Pride,* 5560. Sold at auction, May 8, 1883.

[EAR MARK 80] *Utalpia, 19650.* Solid fawn, black tongue and switch. A very neat, trim animal, below medium size; level, round-barreled, small boned, and light limbed; she has good udder and teats, excellent escutcheon, broad and high.

History—Dropped November, 1879. Bred by Francis le Feuvre, St. Ouen's Parish, Jersey. Sold to E. Burnett for T. A. Havemeyer, and imported in steamship Holland to New York, in August, 1881.

Pedigree—Sire, *Daylight* (P. 282). Dam, *Utopia* (F. 2192).

Progeny at Mountainside—Bull, November 13, 1882, Talpa, 9154, by *Farmer's Pride*, 5560. Sold May, 1883.

Young Mouse, 21493. Solid fawn, shading to gray, light gray head, white tongue, black switch. A handsome cow—level and deep, broad in loin and low in flank, with a beautiful, gracefully-posed and stylish head, small and very rich waxy horns, straight back, thin neck and withers, wide hips; legs straight, tail long and slender; hide mellow, with a very soft, silky coat, yellow skin color, good coupé udder, medium-sized teats.

EAR MARK `81`

History—Dropped July 12, 1879. Bred by Ph. M. le Neveu, St. Clement's. Sold to E. Burnett, for T. A. Havemeyer, and imported in August, 1881, in steamship Holland to New York.

Pedigree

	COLOR.	BREEDER.	SIRE.	COLOR.	SIRE'S BREEDER.
Young Mouse, 21493.... ...	Solid fawn....	Ph. M. le Neveu....	*Landseer* (P. 162)	Gray......	Wm. Luckarift.
Dam, *Mouse* (F. 1787), 11953.	Cho. f'n & gr...	Philip M. le Neveu, St. Clement's, Jersey.			

MOUSE is No. 1 of this Catalogue, and is a remarkably productive and excellent cow.

Progeny at Mountainside—Bull, 1882, *Orator,* 10326, by *Eloquence* (F. 341). Sold in 1883.

EAR 82 MARK *Miss Le Feuvre, 22998.* Solid fawn, shading to dark brown, white tongue, black switch. A superb young cow; head neat, with very light horns (one shelled); very straight backed, wedge-shaped; high, broad and deep behind, with excellent loin and hips, thin neck and withers; udder large, with large teats and milk veins; hide loose and soft, with good color in skin and ears.

History—Dropped January, 1880. Bred by Philip le Feuvre, St. Ouen's, Jersey. Sold to E. Burnett for T. A. Havemeyer, and imported to New York in steamship Holland, in August, 1881.

Pedigree

	COLOR.	BREEDER.	SIRE.	COLOR.	SIRE'S BREEDER.
Miss Le Feuvre. 22998	Solid fawn	Ph. le Feuvre	*Landseer* (P. 162 C.)	Gray	Wm. Luckarift.
Dam, *Blooming* (F. 921 H.C.)	Yellow fawn	P. J. Mourant	—	—	—

Progeny—Bull, October 20, 1882, Mountain Pride, 0000, by *Farmer's Pride*, 5560. Sold May, 1883.

EAR MARK | 83 |

Mahwah Rose, 21708. Silver grey, with dark shadings, white spot on right stifle, black tongue and switch. A most attractive and beautiful animal, and a perfect type of one of the best families of Jersey cows—Nicholas Arthur's "*Rose's.*" She is long bodied, with great depth and width of loin and hips, with straight limbs, a neat head, with dark-fringed ears, a slender horn, rich at base ; a loose, soft, and flexible hide, and a fine, furry coat. She has a well-formed udder, with two extra rudimentary teats ; escutcheon broad flanderine.

History—Dropped May 20, 1881. Bred by Nicholas Arthur, St. Mary's, Jersey. Selected, on the Island, by T. S. Cooper, and imported in August, 1881. Sold with the Cooper-Maddux Herd, in December of the same year, and bought by Mr. Havemeyer.

Pedigree

	COLOR.	BREEDER.	SIRE.	COLOR.	SIRE'S BREEDER.
Mahwah Rose, Island name, *Young Rose. 3d,* 21708......	Silver gray.	Nicholas Arthur.	*Gray of the West* (F. 317 H. C.).	Gray.......	N. Arthur.
Dam, *Rose of Oxford,* Island name, *Rose, 3d* 13469.	Gray.......	Nicholas Arthur.	*Tormentor* (F. 258 H. C.). ...	Gray.......	John Arthur.
2d Dam, *Young Rose* (P. 43 H.C.)	Lt. red & w.	Nicholas Arthur.	*Orange Peel* (F. 129 H. C.)...	Lt. red & w.	John Arthur.
3d Dam. *Rose* (F. 339 H. C.)..	Br'n & white	Nicholas Arthur.	———	———	———

90

EAR 84 MARK **Browse, 14611.** Solid gray fawn, black tongue and switch. A cow above medium size, straight, broad, and deep, with a neat head, short, rich horns, and dished face; body low and capacious; broad and wide hips, drooping rump; legs strong, hide soft and mellow, with furry coat and rich skin color; well-shaped, capacious udder, and large teats.

History—Dropped February 26, 1881. Bred by F. C. Havemeyer, Throgg's Neck, N. Y., and transferred to Mountainside Herd in May, 1882.

Pedigree

	COLOR.	BREEDER.	SIRE.	COLOR.	SIRE'S BREEDER.
Browse, 14611	Solid	F. C. Havemeyer	*Prince Hammond*, 3672	Solid fawn	Fr. le Brocq.
Dam, *Miss Browney*, 7288	Solid	F. le Maistre	*Browney* (P. 158)	Solid lt. br'n.	W. Avril.
2d Dam, *Miss de Cartaret*, 7285	Bronze & w.	F. le Maistre	*Nelson* (P. 65)	Y. f'n & w.	P. du Val.
3d Dam, *Florence*	—	F. le Maistre	—	—	—

EAR MARK	85

Alcmena, 3d, 14614.

Light silver gray, star in forehead, white tongue and mixed switch. An exceedingly beautiful animal, well formed, and of good constitution. She is level, wide loined, broad hipped, and deep in the body, with neat head, dished face, light horns, clean throat, a carcass "hooped" and well ribbed home, deep in flank, with straight legs, good udder and large teats, selvage escutcheon, mellow hide, silky coat, and rich skin color.

History—Dropped May 27, 1881. Bred by F. C. Havemeyer, Throgg's Neck, N. Y. Transferred to T. A Havemeyer, May 3, 1882.

Pedigree

	COLOR.	BREEDER.	SIRE.	COLOR.	SIRE'S BREEDER.
Alcmena, 3d, 14614....	Silver gray..	F. C. Havemeyer..	*Prince Hammond*, 3672....	Solid fawn.	F. le Brocq.
Dam, Alcmena, 6193...	Solid gr. f'n.	W. S. Taylor......	*Nobleman*, 2d(P. 123), 2409.	Solid gr...	J. le Brun.
2d Dam, Corsetta, 1859	Yel. f'n & w.	W. C. Wilson.....	Vice Roy, 346.............	———....	S. J. Sharpless.
3d Dam, Leda, 149....	Lt. f'n & w.	W. C. Wilson.....	Prince, 55...............	Lt. f'n & w.	W. C. Wilson.
4th Dam, *Clara*, 148...	Dk. f'n & w.	J. A. Taintor, impr.	———..................	———....	———

PRINCE HAMMOND is a grandson of *Coomassie*, through *Khedive*, and the excellent qualities of that famous cow are apparent in the well-formed udders, and rich, unctuous hides of his get. *Lille Bonne*, 71 of this herd, the dam of *Prince Hammond*, adds also a rich strain of blood, combining *Queen* (F. 1239), *Plough Boy* (P. 102), and Welcome (F. 207).

NOBLEMAN, 2d, for years led the herd of Wm. S. Taylor, of "The Elms."

Through VICE ROY, 346, Alcmena receives the blood of that noble old cow, *Niobe*, 99, who, in her 18th year, won the special prize of the Am. Jersey Cattle Club, awarded at the Centennial Exhibition.

EAR 86 MARK **Lola Hammond, 12252.** Fawn shading to gray, white on brisket, belly, and flanks; star in forehead, white tongue and mixed switch. A cow of excellent form, grand proportions, and good constitution; head neat, with incurving horns, and full, mild eyes; neck and withers thin; back level and broad; body deep and capacious, with low flanks; hindquarters high, with thin thighs, and long line from hips to rump; legs straight and fine, tail fine; hide mellow and unctuous, with soft, furry coat; skin color creamy; capacious and well-formed udder; escutcheon of the flanderine type.

History—Dropped November 15, 1880. Bred by F. C. Havemeyer, Throgg's Neck, N. Y. Transferred to Mountainside Herd, June 13, 1881.

Pedigree

	COLOR.	BREEDER.	SIRE.	COLOR.	SIRE'S BREEDER.
Lola Hammond, 12252.......	Solid....	F. C. Havemeyer..	Lord Starr, 3746....	Solid dun..	F. R. Starr.
Dam, Lady Hammond, 7286..	Solid....	F. C. Havemeyer..	Sylvester, 2d, 2851..	Fawn & gr.	F. C. Havemeyer.
2d Dam, *Lily Cabot*, 6325.....	Lem. f'n.	J. Cabot..........	——............	——....	——

LORD STARR, a son of Highland Chief, out of Cowles's Eirene, one of the best of the foundation cows of the Echo Farm Herd. HIGHLAND CHIEF is by LITCHFIELD, the Centennial Prize Bull, out of Mell, 4th, a cow of the famous Pansy, 8, family.

LADY HAMMOND is No. 66 of this Catalogue (which see for full pedigree of Sylvester, 2d).

Lille Bonne, 2d, 12253.

EAR MARK: 87

Solid fawn, shading to gray, gray head and forelegs, black tongue and switch. A grand cow—head small, face dishing, eyes full, horns very light and waxy; ears small, dark fringed; back straight and broad, crops full, ribs well sprung, loin very broad; body deep, capacious, and low in the flanks; hips and rump wide and long, legs straight and strong; hide mellow, coat abundant, soft, and furry; color in ears and skin bright orange; udder remarkably well shaped—fore udder superior, teats well set and long; escutcheon first-class "selvage"; naturally carries flesh, and takes it on easily when not milking, but milks down thin.

History—Dropped November 15, 1880. Bred by F. C. Havemeyer, Throgg's Neck, N. Y. Transferred to T. A. Havemeyer, May 3, 1882.

Pedigree

	COLOR.	BREEDER.	SIRE.	COLOR.	SIRE'S BREEDER.
Lille Bonne, 2d, 12253	Solid fawn..	F. C. Havemeyer..	Lord Starr, 3746........	Solid dun	F. R. Starr.
Dam, *Lille Bonne*, 8108....	Solid brown.	F. le Brocq	Plough Boy (P. 102 H.C.)	Light f'n	Wm. Luckarift.
2d Dam, *Queen* (F. 1239)..	———	W. Kerslake.......	———	——— ..	———

LORD STARR'S pedigree in the direct dam's line runs through Cowles' Eirene, Dickenson's and E. Cott's Belle to *Violet*, 23, of Taintor's importation, six generations, the out-crosses being into the best Connecticut and Westchester County herds. On his sire's side he is grandson to LITCHFIELD, the Centennial Prize Bull, and to MEL, 4th, daughter to MEL, 37, by the famous McClellan, 25, and out of a daughter of Norton's *Pansy*, 8, famous in her life, but still more so in her descendants.

LILLE BONNE is No. 71 of this herd, a remarkably handsome, even, excellent, and productive cow, tracing to *Welcome* (207) through *Plough Boy* (P. 102).

EAR MARK 88

Compo, 2d, 14885.
Solid dark gray, black tongue and switch. A medium-sized cow, very elegant, graceful, and stylish, with a beautiful head, well-shaped, up-set horns, and a neck neat and clean as a fawn's; brisket low, belly-line level, barrel round and well-ribbed back; flanks deep; hindquarters fine and square; straight, deer-like legs; mellow hide, with silky coat and creamy color in ears and skin; well-shaped udder and teats.

History—Dropped September 19, 1881. Bred by Wm. Simpson. Bought in dam at Simpson's Sale of May, 1881, by F. C. Havemeyer, and transferred to T. A. Havemeyer, May 3, 1882.

Pedigree

	COLOR.	BREEDER.	SIRE.	COLOR.	SIRE'S BREEDER.
Compo, 2d, 14885....	Solid fawn.	Wm. Simpson....	Invincible, 5510....	Solid, bl'k pt..	Wm. Frazer, Ky.
Dam, Compo, 11844..	Solid......	Wm. Simpson....	*Councillor*, 4468....	Solid fawn.....	Wm. Amy, St. Peter's.
2d Dam, Silene, 4307.	Solid fawn.	Wm. Simpson....	*St. Helier*, 45......	Silver gray.....	O. S. Hubbell, impr.
3d Dam, Lara, 4306..	Dun & w...	O. S. Hubbell.....	Arab, 245..........	Fawn	L. H. Twaddell.
4th Dam, Effie, 523..	Fawn.. ...	John Glenn......	*Commodore*, 56......	Fawn & white..	John A. Taintor, impr.
5th Dam, *Countess*, 897	———	John Glenn, impr.			

INVINCIBLE is by Thorndale, 2582, out of Phædra, 2561, a "pure Alphea" cow, bred by Col. Hoe, by Mercury, 432, out of Leda, 799, a daughter of Europa, 176. Invincible stood at the head of the Gold Medal Herd, and won himself *first* as a two-year-old at the Elmira Show of the N. Y. State Agricultural Society in 1881.

COUNCILLOR, 4468, is a great-grandson of *Coomassie* (F. 1442 H. C.), through *Tormentor* (F. 258 H. C.), and *Khedive* (P. 102 H. C.).

ST. HELIER, 45, was possessed of astonishing prepotency, and was for years the handsomest and best bull in New England in the judgment of many excellent breeders.

PHÆDRA, 2561, is the best daughter of Mercury, having a record of 19 lbs. 12 oz. of butter in 7 days, by Mr. Simpson's test.

Lady Pert, 14613. Solid fawn, gray head, black tongue and switch. A cow of excellent form, medium size, fine bone, good constitution, and great style. She is fine in head and horn, with a capacious carcass, straight, broad back, deep flanks, good hindquarters, and fine, deer-like legs ; hide mellow and soft, coat furry, skin color creamy, udder and teats excellent—one extra rudimentary teat ; potdevin escutcheon.

EAR MARK: 89

History—Dropped April 10, 1881. Bred by J. A. Desreaux, Perry Farm, St. Mary's Parish, Jersey. Imported in dam, October, 1880, by A. A. Degrauw, by steamer Gloucester to New York. Sold to F. C. Have-meyer in 1881, and transferred to Mountainside Farm, May, 1882.

Pedigree

	COLOR.	BREEDER.	SIRE.	COLOR.	SIRE'S BREEDER.
Lady Pert, 14613............	Solid fawn....	J. A. Desreaux......	King (P. 238)...	Light brown.	F. le Brocq, Jr.
Dam, *Lady Perry*, 11908.....	Fawn.........	J. A. Desreaux......	———.........	———......	———

EAR `90` **MARK**

Fairy Golddust, 14612.
Solid rich brown, with gray head, black tongue and switch. A beautiful creature, of medium size, nearly perfect form, and great quality. She has a light, graceful head, with very rich, neat horns, a thin, clean neck, straight, broad back, deep flank, low brisket and belly. She has superb hindquarters, a mellow hide, soft furry coat, and rich orange skin; udder of good form and size; teats well placed, one rudimentary; escutcheon flanderine.

History—Dropped March 20, 1881. Bred by H. C. Kretschmar. Bought, in her dam, at the Kellogg Combination Sale, October 13, 1880, and dropped upon the farm of F. C. Havemeyer, at Throgg's Neck, and transferred to the Mountainside Herd with her dam, May, 1882.

Pedigree

	COLOR.	BREEDER.	SIRE.	COLOR.	SIRE'S BREEDER.
Fairy Golddust, 14612.....	Solid f'n..	H. C. Kretschmar......	*Jersey Golddust*, 2134..	Cream....	Wm. Amy.
Dam, Annie Golddust, 6849.	Solid f'n..	H. S. Parke..........	*Jersey Golddust*, 2134..	Cream....	Wm. Amy.
2d Dam, Annie Page, 2690.	Solid f'n..	H. S. Parke..........	Son of Alphea, 562....	Sold f'n...	R. M. Hoe.
3d Dam, *Princess*, 1154...	——...	W. H. T. Hughes, impr.			

JERSEY GOLDDUST was selected on the Island to fill a special order, by M. C. Weld, for H. S. Parke. He proved to be a sire of great prepotency, impressing his own characteristics upon his offspring for several generations. He was famous for extraordinary richness of skin, excellence of form, and for the uniform beauty and high quality of his get.

SON OF ALPHEA was the last calf of his famous dam. He was by *Dolphin* 2d, and so introduces a strong dash of "Dauncy" blood, which has proved in numerous instances a cross of remarkable merit.

PRINCESS was an imported cow of extraordinary grace and beauty, as well as richness and quantity of yield.

ANNIE PAGE dropped her first calf at the age of twenty months, but ten months *previous* was found with an udder distended with milk, and after that was milked regularly and dried off before calving, like any old cow.

FAIRY GOLDDUST'S blood is 75 per cent. *Jersey Golddust*, 12½ *Princess*, 6¼ Alphea, and 6¼ "Dauncy."

EAR 91 MARK

Kate le Brocq, 14615.
Fawn, nearly solid, star, flecks of white on belly, black tongue and switch. A superb cow of great quality and richness; level, broad, and deep, with a neat head, rich horn, straight, deer-like limbs, slender tail, and stylish carriage. Her hide is extremely loose and mellow, with a long, silky coat and very rich skin color. She has an udder of excellent form, medium-sized teats, and a selvage escutcheon, well out upon the thighs.

History—Dropped October 11, 1881. Bred by F. C. Havemeyer, Throgg's Neck, N. Y. Transferred to T. A. Havemeyer, May 3, 1882.

Pedigree

	COLOR.	BREEDER.	SIRE.	COLOR.	SIRE'S BREEDER.
Kate le Brocq, 14615.......	Fawn....	F. C. Havemeyer.. ...	Lord Starr, 3749..	Solid dun...	F. R. Starr.
Dam, *Nellie le Brocq*, 8986..	Bronze...	Wm. Holt, impr.......	———.........	———	———

LORD STAR is a grandson of LITCHFIELD, the "Centennial Prize Bull," through Highland Chief, whose dam, Mel, 4th, daughter of MEL, 37, of the famous *PANSY*, 8, family, goes on her sire's side to *Lily*, 1, a most excellent cow, and is a double granddaughter to McCLELLAN, 25, a combination of blood of exceeding richness and promise. On the side of his dam, Cowles' Eirene 2d, Lord Starr traces his descent to *Bronx, Bashan, Lady Webster*, and other famous Westchester County names, and, besides, to a long line of worthies of the Taintor, Norton, and Buck importations.

Beacon Lass, 14834. Solid gray fawn, black tongue and switch. A large cow, well shaped, deep bodied, level, and broad; horns light, ears dark-fringed and rich in color; back straight, brisket and belly deep; body capacious, with well-sprung ribs; flank low, rump slightly depressed, thighs long and flat, legs straight, and tail fine; hide mellow, coat furry, very rich in color, and unctuous; udder and teats large; escutcheon broad, well out on thighs.

[EAR MARK: 92]

History—Dropped January 5, 1881. Bred by F. C. Havemeyer, Throgg's Neck. Transferred to T. A. Havemeyer, May 3, 1882.

Pedigree

	COLOR.	BREEDER.	SIRE.	COLOR.	SIRE'S BREEDER.
Beacon Lass, 14884.........	Solid fawn.	F. C. Havemeyer.	*Prince Hammond*, 3672	Solid fawn..	F. le Brocq.
Dam, Josephine Beacon, 3306	Dark fawn.	Beacon Farm.....	Touchstone, 315......	——	T. J. Hand.
2d Dam, Josephine, 2d. 3296.	Smoky f'n.	Beacon Farm.....	Beacon Comet, 675....	Dappled f'n.	R. L. Maitland.
3d Dam, *Josephine*, 2737....	Mul. fawn.	Eli Hubert—M. H. Cochrane, impr.			

The fame of the Beacon Farm Herd of Jerseys, which was great and merited, depended largely upon that magnificent cow, "Old" JOSEPHINE, 2737, her daughter, JOSEPHINE, 2d, and the famous bull BEACON COMET, 675, all of whom occur in this pedigree. The last named was by Maitland's *Comet*, who was imported in his dam, *English Beauty*. On his (Beacon Comet's) dam's side he traces to Captain Darling, by *Prince of Jersey*, and out of *Jersey Queen*, and to *Belle*, 225, a grand cow of the Paterson-Colt herd.

TOUCHSTONE, on the other hand, traces to the Hartford-Colt early importations, and to those of Norton and Taintor, including in his pedigree, besides Norton's bull, *Splendid*, 2, the three famous Fannies, Affleck's No. 21, Colt's No. 44, and Beach's No. 72.

PRINCE HAMMOND grafts upon this grand old stock the freshly imported blood of *Lille Bonne*, No. 71 of this herd, and of *COOMASSIE*, through *Khedive*.

Fancy Fan, 12657. Solid brown, gray head, full black points, tongue, and switch. A noble cow, of nearly perfect frame, and of great beauty and style; deep and rich milker. She has a handsome head, with incurving, waxy horns; neck and withers clean and thin; back straight to hips, and rump higher than hips; thighs long and flat, and rump long; legs small boned, and straight; tail long and slender; hide mellow, thin, and loose, with fine, silky coat and creamy skin color; udder very large and well shaped, with large teats; milk-veins large and tortuous; and a selvage escutcheon of good breadth.

EAR 93 MARK

History—Dropped April, 1875. Bred by Eli Hubert, St. Ouen's Parish, Jersey. Imported by John D. Wing, in 1881, in steamship Marengo to New York, and bought by T. A. Havemeyer, at the Kellogg Combination Sale, in May, 1882.

Pedigree

FOLLOWING THE SIRE'S LINE.	COLOR.	BREEDER.	DAM.	COLOR.	DAM'S BREEDER.
Fancy Fan, 12657...	Solid brown...	Eli Hubert......	*Fanny* (F. 1150)...	Brown...	T. Wills.
Sire, *Loyal* (P. 70 H. C.).....	Brown........	John le Brun....	*Stella* (F. 705).....	Brown...	John le Brocq.
2d Sire, *Yankee* (P. 27 H. C.).	Light brown...	E. Gibaut.......	*Georgette* (F. 309)..	Fawn....	E. Gibaut.
3d Sire, *Paddy* (F. 97)........	Brown & w....	E. Gibaut.......	———	———...	———

YANKEE won *first* over all Jersey in 1871, *first,* Herdbook Society's prize, and also the Silver Medal at the Channel Island Exhibition the same year. His sire, *Paddy,* was also a prize-winner.

FANCY FAN, aside from her intrinsic excellence, gained her chief fame from the fact that she was a close competitor of that remarkable cow *Coomassie*—the two cows having "scaled" the same number of points each. She suffered from the sea-voyage, from acclimating, and from being "put in condition" for sale, all within a few months, but is now in excellent form, and improving steadily both in appearance and usefulness.

[EAR MARK 94] *Stolen Kisses (F. 3247), 16864.* Solid bronze fawn, black-fringed ears, full black points, tongue, and switch. A medium-sized cow of great beauty, wedge shaped, straight, and fine; head slender, dished, with large, prominent eyes, incurving horns; neck and withers thin; back straight to hips, with broad loin and slightly drooping rump; body capacious, well-ribbed back; hide mellow, coat silky, and skin rich orange; excellent udder and teats; broad limosine escutcheon.

History—Dropped May, 1880. Bred by J. P. Bosdet, St. Brelades' Parish, Jersey. Imported by John D. Wing, 1882. Sold at the Kellogg Combination Sale, May, 1883, and bought by T. A. Havemeyer.

Pedigree—*Stolen Kisses*, being of the Foundation Stock of the Island, has no recorded pedigree.

Maid of the Valley, 16556.

EAR MARK 95

Nearly solid orange fawn, with white tongue and switch. A very handsome cow, about medium size, fine, level, long bodied, close ribbed, deep, and low flanked. She has large udder and teats, strong milk-veins, rich, mellow hide, a deep skin color, a broad flanderine escutcheon, and a grand constitution.

History—Dropped May 10, 1878. Bred by W. W. Wheaton, Binghamton, N. Y. Bought by J. V. N. Willis, May 18, 1882, and sold by him at the Kellogg Combination Sale, May, 1883, and bought by T. A. Havemeyer.

Pedigree

	COLOR.	BREEDER.	SIRE.	COLOR.	SIRE'S BREEDER.
Maid of the Valley, 16556...	Lemon fawn.	W. W. Wheaton...	Vernon, 1071.......	F'n & w.	T. J. Hand.
Dam, Arlinda, 14780.......	French gray..	W. W. Wheaton...	Rubicon, 1202......	——..	Jos. Juliand.
2d Dam, Constantine, 1634..	Fawn & w...	A. A. Underhill.. .	Capt. Jinks, 645....	F'n & w.	J. D. Wing.
3d Dam, Constance, 1633...	Fawn & w...	W. H. Aspenwall..	Young Taurus, 387..	Black ...	W. H. Aspenwall.
4th Dam, Daphne, 584......	Fawn & w...	W. H. Aspenwall..	Taurus, 244........	Fawn ...	W. H. Aspenwall.
5th Dam, *Iris*, 89..........	Cream......	W. H. Aspenwall, impr.			

VERNON was by Marius, 760 (double *Lady Mary*), out of *Velvet*, 294, an imported cow (dam of *Tancred*, 501), thus half-brother to SIGNAL, a famous getter of butter cows, including Tennella, credited with 22 lbs. 1½ oz., Croton Maid, 21 lbs. 11½ oz., Belle of Paterson, 16 lbs. 6 oz., Valhalla, 16 lbs. of butter in 7 days.

MAID OF THE VALLEY won *first* at the New Jersey State Show in 1882, also *first* at Freehold and at Mount Holly, N. J., the same year.

EAR `96` **MARK**

Carlotta, 18533. Buffalo fawn—white star in forehead, white on belly, thighs, and hind legs. A fine, large cow, straight backed to the tail, broad in loin and hips, with a capacious carcass—very deep; limbs and tail light; udder capacious, with large milk-veins; hide with silky coat and rich skin color; escutcheon broad bi-corn.

History—Dropped June, 1879. Bred by Edward Denize, St. Lawrence Parish, Jersey. Imported by John F. Maxfield to New York in December, 1882. Sold, at the Kellogg Combination Sale of May, 1883, to T. A. Havemeyer.

Pedigree

FOLLOWING THE LINE OF SIRES.	COLOR.	BREEDER.	DAM.	COLOR.	DAM'S BREEDER.
Carlotta, 18533	Fawn	Edward Denize.	*La Belle*	—.	Edward Denize.
Sire, *Carlo* (P. 180), 5559	Orange f'n.	Thos. Falla, Jr.	*Pretty Maid* (F. 1493)	Brown.	Thomas Falla, Jr.
2d Sire, *Hero* (P. 126 H. C.)	Gray	Thos. Falla, Jr.	*Cowslip* (P. 24)	Brown.	T. Falla.
3d Sire, *Dick* (F. 171 H. C.)	Lt. brown.	Thos. Falla.	*Cherry* (F. 1140)	Lt. red.	T. Falla.
4th Sire, *Yankee* (P. 27 H. C.)	Lt. brown.	E. Gibaut.	Georgette (F. 309)	Red.	E. Gibaut.
5th Sire, *Paddy*	Lt. br. & w.	E. Gibaut.	——	—.	——

EAR MARK | 97 |

Lilly of Cedar Grove, 13912.

Cream fawn in summer, light cinnamon in winter; white tongue, dark switch. A large, noble-looking cow, formed for a deep milker; head rather long, dished, with slender, incurving horns; neck light, withers thin, back straight and level to hips; body deep and capacious, with well-sprung ribs, broad loin, and deep flank; hide thin, with soft silky coat; legs and tail slender; excellent, well-quartered udder, full milk-veins, and a left-hand flanderine escutcheon.

History—Dropped November 27, 1878. Bred by P. W. Myer. Sold at Kellogg's Combination Sale in May 1883, and bought by T. A. Havemeyer.

Pedigree

	COLOR.	BREEDER.	SIRE.	COLOR.	SIRE'S BREEDER.
Lilly of Cedar Grove, 13912......	Solid fawn..	P. W. Myer.	Cedar Grove Lad, 6409.	Solid d'k f'n..	P. W. Myer.
Dam, Bessie of Cedar Grove, 13907.	Sol. or. f'n..	P. W. Myer.	Rover, 2278...........	Gr. f'n & w...	P. W. Myer.
2d Dam, *Quill*, 1103	Buff fawn...	Edward Howe, imp'r.			

LILLY'S dam and grand-dam were both excellent butter cows—the former making 12 lbs. on pasture alone.

EAR MARK |98|

Colt's La Biche, 6399.
Solid smoky fawn, black points, tongue, and switch. A noble, large-sized, handsome cow, with a well-shaped head, having spreading, rich-colored, up-set horns, thin withers, level back, very broad in loin and hips; a deep, capacious body, low flank, high and broad hindquarters, long, flat thighs, grand udder and teats; mellow, soft, silky hide, with rich orange color in ears, large, tortuous milk-veins, and broad flanderine escutcheon; excellent constitution. .

History—Dropped June 21, 1877. Bred by Sam. C. Colt. Entered by D. A. Givens, in the Kellogg Combination Sale of May, 1883, and bought by T. A. Havemeyer.

Pedigree

	COLOR.	BREEDER.	SIRE.	COLOR.	SIRE'S BREEDER.
Colt's la Biche, 6399	Smoky fawn..	S. C. Colt..	Knave, 1856	Solid red fawn.	J. B. Williams.
Dam, La Biche, 2d, 4023	Light fawn...	S. C. Colt..	*Pierrott*, 636	Silver fawn ...	S. C. Colt, impr.
2d Dam, *La Biche*, 905	Fawn	S. C. Colt, impr.			

KNAVE'S dam was DUSKY, a granddaughter of ALBERT, 44, who is reckoned as one of the best "butter bulls" of the country.

PIERROTT, the sire of La Biche, 2d, won great fame for himself and his owner's herd as the sire of many rich and beautiful cows. The statement, therefore, is not surprising that she made, in April, 1882, 16 lbs. $14\frac{1}{2}$ oz. of butter in 7 days; and again, that in 1883, on dry feed only, in 31 days, commencing February 14th, she gave 73 lbs. 2 oz. of butter, the score for the first 14 days being 34 lbs. 5 oz., or 17 lbs. $2\frac{1}{2}$ oz. per week.

EAR
99
MARK

Island Flower, 15265. Solid brown, shading to gray, with black tongue and mixed switch. A fine-looking, medium-sized cow, wedge shaped, with thin neck and withers, broad loin and hips, neat, deer-like legs, and thin tail; slightly-drooping rump, capacious barrel, deep in brisket and flank; capacious udder—excellent fore-udder; hide thin, with silky coat; curveline escutcheon.

History—Dropped January, 1880. Bred by John Pinel, St. John's Parish, Jersey. Imported by William Simpson in steamer Marengo, January 14, 1882. Sold at his annual sale in May of the same year. Bought by Wm. Watson, and, after remaining in his herd a year, was sold by him, with the remainder of his herd, May, 1883, and bought by T. A. Havemeyer.

Pedigree

	COLOR.	BREEDER.	SIRE.	SIRE'S COLOR.	SIRE'S BREEDER.
Island Flower, 15265....	Solid Brown...	John Pinel....	*Carlo* (P. 180)..	Orange fawn....	Thos. Falla, Jr.
Dam, *Actress* (F. 1615)..	Solid Brown....	P. Pinel......	———	———	———

Nymph of St. Lambert, 12968.

EAR MARK: 100

Nearly solid dark fawn, dark-fringed ears, black tongue and switch. A handsome, medium-sized, wedge-shaped cow, with a neat head, light horn, thin neck and withers; level to hips, with a slightly sloping rump, a broad loin, thick thighs, close twist, capacious body, with low brisket and belly line; hide mellow and rich, with silky, unctuous coat; excellent color in ears; capacious udder, with good-sized teats and prominent milk-veins.

History—Dropped May 3, 1880. Bred by R. H. Stephens, St. Lambert, P. Q., Canada. Sold to Valancey E. Fuller, December 12, 1881, and by him entered in the Kellogg Combination Sale of May, 1883, and bought by T. A. Havemeyer. Just previous to the sale, Nymph, by owner's test, gave 6 lbs. 9 oz. in $3\frac{1}{2}$ days, equivalent to 13 lbs. 2 oz. in a week.

Pedigree

	COLOR.	BREEDER.	SIRE.	COLOR.	SIRE'S BREEDER.
Nymph of St. Lambert, 12968..	Sol. d'k f'n..	R. H. Stephens..	Stoke Pogis, 3d, 2238.	Mul. f'n.	P. le Clair.
Dam, Diana of St. L't, 6636 ...	Solid fawn ..	R. H. Stephens..	Stoke Pogis, 3d, 2238.	Mul. f'n.	P. le Clair.
2d Dam, Pet of St. L't, 5123...	Fawn.	R. H. Stephens..	Lord Lisgar, 1066....	Solid....	S. S. Stephens.
3d Dam, Lucy of St. L't, 5116..	Gray	R. H. Stephens..	*Victor Hugo*, 197.....	Bl'k & b'n	S. S. Stephens, impr.
4th Dam, *Lydie*, 495..........	——.....	Mr. Le Templier—S. Sheldon Stephens, impr.			
5th Dam, *Bonnie*, 491........	Brown......	Mr. Le Templier—S. Sheldon Stephens, impr.			

The strong infusion of Dauncy blood received from STOKE POGIS, 3d, whose dam was *Marjoram*, 3229, a 16-lb. cow, imported by Peter le Clair, and into which the blood of *Rioter* (746 E. H. B.), the sire of Hoe's *Rioter*, 2d (the sire of Eurotas and Black Prince of Hanover), largely enters, makes this a pedigree of great promise, especially as other daughters of Stoke Pogis, 3d, have proved excellent butter cows. First among them being Mary Ann of St. Lambert, who by Mr. Fuller's test gave in 62 days (June and July, 1883) 209 lbs. 2 oz. of salted butter, the first 31 days giving 106 lbs. 12 oz., and the best week 24 lbs. 13 oz.

PET OF ST. LAMBERT is inbred to *Victor Hugo*, both sire and dam being sired by him.

EAR
| IOI |
MARK

Casta Cazique, 16963.

Chocolate brown, with white on belly and hind legs, black and white tongue, and mixed switch. A cow below medium size, neat and well formed, with a good head, a rather strong, short horn and fringed ears; neat, clean neck, thin withers, level to hips; rump falling off; fine, deer-like limbs; mellow hide, wiry coat, rich orange skin-color; well-shaped, good-sized udder and teats, one extra rudimentary teat, and a potdevin escutcheon, of good breadth. A hardy, strong-constitutioned animal.

History—Dropped June 3, 1881. Bred by J. R. Eoff. Entered by D. A. Givens in the Kellogg Combination Sale of May, 1883.

Pedigree

	COLOR.	BREEDER.	SIRE.	COLOR.	SIRE'S BREEDER.
Casta Cazique, 16963..........	Gr. fawn...	J. R. Eoff......	Cazique, 3103......	Fawn...	W. L. & W. Rutherford.
Dam, Casta Diva, 8154........	Gr. f'n & w.	Belpré Stock Co.	Pope, 738	Solid....	John L. Stettinius.
2d Dam, Belpré Belle, 2d, 7390.	Fawn & w..	Belpré Stock Co.	*Crown Prince*, 581..	F'n & w.	Jean Baudiens.
3d Dam, *Belpré Belle*, 1488	———.....	John Picot.	B. Kittredge, imp'r.		

EAR MARK 102 Compo, 11844. Solid buffalo fawn, black tongue and switch. A very neat, close-knit, deer-like cow; level, compact, round barrelled, close ribbed, broad chested, low flanked, and fine limbed; excellent udder, full milk veins, and flanderine escutcheon.

History—Dropped March 31, 1880. Bred by Wm. Simpson. Sold to F. C. Havemeyer in 1881. Dropped her first calf, Compo, 2d, No. 88 of this Catalogue, at 17½ months old, was entered in the Kellogg Combination Sale of May, 1883, and bought by T. A. Havemeyer.

Pedigree

	COLOR.	BREEDER.	SIRE.	COLOR.	SIRE'S BREEDER.
Compo, 11844	Solid buffalo	Wm. Simpson	*Councillor*, 4468	Solid fawn	Wm. Amy.
Dam, Silene, 4307	Gray fawn	O. S. Hubbell	*St. Helier*, 45	Salmon & gr.	Philip Quenault.
2d Dam, Lara, 4306	Dun & white	O. S. Hubbell	Arab, 245	Fawn	L. H. Twaddell.
3d Dam, Effie, 523	Fawn	John Glenn	*Commodore*, 56	Lt. f'n & w.	John A. Taintor, impr.
4th Dam, *Countess*, 897	——	John A. Taintor, impr.			

SILENE is a "fourteen-pound" cow, and ranks well among the many superior daughters of *St. Helier*—one of the most excellent bulls of New England both in appearance and the butter quality of his get.

COUNCILLOR, at the head of Maj. Campbell Brown's Herd, is a great-grandson of *Coomassie*, through *Tormentor* and *Khedive*, and makes a grand "top-cross" upon this excellent old-fashioned pedigree.

EAR 103 MARK

Jersey Rose (F. 3633 H. C.), 18216. Fawn and white—the white confined to belly, flanks, and extremities—and fleck on left shoulder; white tongue and switch; head and neck gray. She is a medium-sized cow, of neat, compact build and fine bone; head neat, with very rich horn; neck and withers thin; back straight to rump, which is slightly drooping; loin broad; ribs well sprung; skin mellow, with soft, silky coat; udder capacious, with large teats; selvedge escutcheon.

History—Dropped September, 1880. Bred by John Picot, Trinity Parish, Jersey. Imported by T. S. Cooper in steamer Greece, November 20, 1882. Sold at auction in New York, May, 1883, and bought by T. A. Havemeyer.

Pedigree

	COLOR.	BREEDER.	SIRE.
Jersey Rose, 18216 Island name, *Totness* (F. 3633 H. C.)	Fawn & white..	John Picot..	*Nicolls's bull*—not presented for Herdbook examination.
Dam, *Guenon's Pride* (F. 2767 C.) ——......		John Picot..	——.

Mr. Cooper states that Guenon's Pride was a 16 quart cow, and made 9 pounds, Jersey weight, of butter three months after calving, and that her dam was a grand butter cow; also that her sire's dam made 11½ lbs. in seven days. The Jersey pound is about one-tenth heavier than ours.

EAR |104| **MARK**

Cream of Jersey (F. 2768), 18181.

Lemon fawn, star in forehead, white on belly, flanks, and thighs, besides scattered flecks; white tongue, mixed switch. A fine, large cow of the Coomassie type—straight, wedge shaped, broad and low; head well shaped, with incurving horns, black tipped, rich, and waxy; neck fine and thin; rump high, tail and limbs fine; hide thin, with fine coat; udder of excellent form, well quartered and capacious, with large teats; milk-veins prominent and tortuous; escutcheon left-flanderine of the first class.

History—Dropped in 1878. Bred by John Picot, Trinity Parish, Jersey. Qualified for admission to the Herdbook, May, 1881. Imported by T. S. Cooper, November 20, 1882, in steamship Greece to New York. Sold at auction in New York, May, 1883, and bought by T. A. Havemeyer.

Pedigree

FOLLOWING THE SIRE'S LINE.	COLOR.	BREEDER.	DAM.	COLOR.	DAM'S BREEDER.
Cream of Jersey (F. 2768)...	Lem. f'n & w.	John Picot.....	*Jessy* (F. 878)............	Br'n & w.	John Picot.
Sire, *Guy Fawkes* (F. 251 H. C.).	Light gray...	Philip Godeaux.	*Angelica* (F. 1738 H. C.)..	Fawn...	C. Larbalestier.
2d Sire, *Koffee* (F. 233 H. C.)....	Dark brown..	Philip Godeaux.	*Coomassie* (F. 1442 H. C.)..	Brown ..	C. F. Dorey.
3d Sire, *Nonpareil* (P. 37 H. C)..	Red & white..	J. P. Nicholls..	*Les Catiaux* (F. 487 H. C.).	Red & w.	Alfred le Gros.
4th Sire, *Orangepeel* (F. 129 H. C.).	Lt. red & w..	J. Arthur......	*Cowslip* (F. 330 H. C.)....	Brown ..	J. Arthur.
5th Sire, *Clement* (F. 61 H. C.)...	Lt. red & w..	E. Gibaut......	*Clementine* (F. 232).......	Lt. br'n.	E. Gibaut.
6th Sire, *Willie* (F. 12 H. C.)....	Gray & white.	Jas. Ahier......	——...............	——..	——

GUY FAWKES is a grandson of *Coomassie* and of *Garenne*, both cows of great fame.

The whole connection on the sire's side abounds in famous animals, both as great producers and as prize winners.

JESSY, Cream's dam, is represented as an 18-quart cow.

EAR
| 105 |
MARK

Mary Jane (F. 3115 C.), 18185. Yellow fawn, with white on belly and fleck on right stifle; white tongue, black switch. A very handsome, large cow, with a beautiful head, dished, marked with dark lines, having light, rich, incurving horns and full, dark-rimmed eyes; neck fine and clean; withers thin; back straight, with broad loin; capacious body, well-ribbed back, and deep in brisket and girth; tail and limbs fine; hide mellow, with creamy skin color; udder large and handsome, with large teats; broad selvage escutcheon, and prominent milk-veins.

History—Dropped in 1878. Bred by Philip Gibaut, St. John's Parish, Jersey. Imported by T. S. Cooper, in steamer Greece, November, 1882. Sold at auction in New York, May, 1883, and bought by T. A. Havemeyer.

Pedigree—Being of the *Foundation Stock* of the Island, *Mary Jane* has no recorded pedigree.

| EAR 106 MARK | *Golden Touch* (*F. 3509 C.*), *18221*. Fawn and white, the white chiefly upon belly and legs; star, spot on withers and at base of tail; white tongue and switch. An exceedingly graceful and beautiful cow; head strikingly fine, with slender, down-curving, very yellow horns; neck and withers thin; back straight; body wedge shaped, deep at flank and broad in the loin and hips, with fine limbs and tail; hide thin, with fine, silky coat and creamy skin color; udder large, soft, well quartered, with long, slender teats and strong milk-veins; escutcheon flanderine.

History—Dropped September, 1880. Bred by John Arthur, of La Pompe, St. Mary's Parish, Jersey. Qualified for admission to the Island Herdbook, May, 1881. Imported by T. S. Cooper, in steamship Greece, to New York, November, 1882. Sold at auction, May, 1883, and bought by T. A. Havemeyer.

Pedigree—Being of the *Foundation Stock* of the Island, *Golden Touch* has no recorded pedigree.

EAR MARK 107

Deerfoot Girl, 15329.

Solid fawn, full black points, tongue and switch. A grand cow, of large size, excellent form, and every evidence of strong constitution. Her head is well shaped, with a broad muzzle, broad, distinct fillet, full, mild eyes, horns of exceptionally beautiful form—flat, incurving, rich at base—and with dark-fringed ears; her neck light and clean; withers thin; back straight and level to setting on of tail; ribs well sprung, body capacious; hindquarters excellent, thighs long and broad, legs fine and straight; hide mellow and rich, orange in ears; udder capacious, of excellent form, with large tapering teats; milk-veins prominent and tortuous; escutcheon first-class selvedge.

History—Dropped November 18, 1877. Bred by Edward Burnett, Deerfoot Farm. Sold by him to T. A. Havemeyer in July, 1883. By Mr. Burnett's test, in November, 1882, she yielded 15½ lbs. of butter in seven days.

Pedigree

	COLOR.	BREEDER.	SIRE.	SIRE'S COLOR.	SIRE'S BREEDER.
Deerfoot Girl, 15329.............	Sol. f'n.	E. Burnett....	{ *Sweepstakes Duke*, 1905 } { Island name, *Duke* (P. 76). }	Gray.....	C. Lesbirel.
Dam, Deerfoot Maid, 5701......	Sol. f'n.	E. Burnett....	*Albion*, 490.............	Bl'k & gr.	C. Wellington, imp'r.
2d Dam, Daisy of Deerfoot, 3182.	Fawn ..	Jos. Burnett ..	*Czar*, 273	————...	J. A. Taintor, imp'r.
3d Dam, *Fanny*, 675...........	Mouse .	Jos. Burnett, imp'r.			

DUKE, 76, was imported the autumn after he was a year old, yet he left notable progeny on the Island—*Gray King* (the sire of *Farmer's Glory*), *Vertumnus* (sire of *Punchinello*, *Day Dream*, *Tidy*, 2d, etc.—grand cows), and *Royalist* (sire of *Lady Arthur* and *Royalist's Gypsy* of this herd) are, like their sire, all first-prize winners over the whole Island.

DAISY OF DEERFOOT won 1st in her class at the New York State Fair in 1873.

EAR MARK |108|

Polly of Deerfoot, 15328.
Silver fawn, with white fleck on each shoulder, white tongue and switch. A large, well-formed cow. She has a long head, with large, mild, dark-rimmed eyes, downward, incurving, rich horns; a clean neck, thin withers, straight back, drooping at the rump; a broad loin, flat ribs, capacious carcass, and low flank; her legs are straight and strong, her hide mellow, coat rather harsh but unctuous, skin color rich; udder large; teats large and tapering; milk-veins prominent and well expanded.

History—Dropped July 25, 1877. Bred by E. Burnett, Deerfoot Farm. Sold to T. A. Havemeyer, July, 1883. By Mr. Burnett's test she made 15 lbs. 2 oz. of butter in February, 1882.

Pedigree

	COLOR.	BREEDER.	SIRE.	SIRE'S COLOR.	SIRE'S BREEDER.
Polly of Deerfoot, 15328	Silver fawn	E. Burnett	Deerfoot Boy, 1926	Solid gray	E. Burnett.
Dam, Patty of Deerfoot, 15321	Solid red fawn	E. Burnett	*Albion*, 490	Black & gray	C. Wellington, imp'r.
2d Dam, Pink, 4, 3983	White & fawn	Jos. Burnett	Jersey Boy, 272	F'n & white	Jos. Burnett.
3d Dam, *Pink*, 676	———	Jos. Burnett, imp'r.			

This pedigree shows some close breeding, of which the apparent results are very good. The cow's sire and dam are both by *Albion*, 490, and JERSEY BOY and DAISY OF DEERFOOT, the dam of DEERFOOT BOY, are full brother and sister.

FAMOUS ANIMALS

REPRESENTED BY PROGENY IN THIS CATALOGUE.

	PAGE		PAGE		PAGE
Albert, 44	39, 40, 66, 67	*Clement* (F. 61), 115	40, 41, 70, 111	*Europa*, 176	12, 13, 68, 69
Alphea, 171	13, 68, 69, 78, 95, 97	*Coomassie* (F. 1442)	29, 30, 34, 90, 91, 95	Eurotas, 2454	12, 13, 69
Angelina Baker, 13	42, 64		99, 109, 111	*Farmer's Glory* (F. 274), 5196	25, 27
Artist (F. 276)	86	*Councillor*, 4468	95, 109		43, 80
Barney, 1491	37, 60	*Czar*, 273	56, 63, 114, 115	*Favorite of the Elms*, 1656	67, 73, 76
Beacon Comet, 675	99	*Dandelion*, 2521	67	*Flora*, 113, and *Countess*, 114 (Motley's)	11
Belle Dame, 11951	12	*Daylight* (P. 283)	87		56, 63
Black Prince of Hanover, 2873	13	*Dauncy Herd*	12, 13, 69, 74, 78, 97, 107	*Garenne* (F. 1575)	30, 111
Bonheur (F. 1651)	8, 25, 27, 43, 46, 80	*Dolphin*, 2d, 468	13, 69, 74, 78, 97	*Grey King* (P. 169)	8, 9, 25, 27, 31, 43, 80
Bronx, 306	98	Duke of Darlington, 2460	12, 13, 69	*Grey Prince* (F. 168)	19, 26, 27
Brownie (P.158)	20, 21, 22, 38, 46, 73, 81, 91	*Duke* (P. 76), 1905	8, 9, 25, 27, 28, 43, 52	Guy Fawkes (F. 251)	30, 111
Carlo (P. 180), 5559	32, 103, 106		80, 114	*Hero* (P. 90)	24, 25, 48, 50
Cherry (F. 1140)	10, 32, 33, 103	*Edith*, 167	42, 65	*Hero* (P. 126)	32, 103

	PAGE		PAGE		PAGE
Jersey Belle of Scituate, 7828	11	*Nobleman*, 2d (P. 123), 2409	72, 76, 92	Silene, 4307	95, 109
Jersey Golddust, 2134	78, 97	Orange Peel (F. 129)	90	*Splendid*, 2	39, 40, 61, 64, 67, 99
Josephine, 2737	99	Phædra, 2561	95	*St. Helier*, 45	59, 66, 95, 109
Jupiter, 93 (see Alphea, also)	42	Pansy, 8	39, 40, 59, 64, 66, 67, 94, 98	*St. Martin*, 1482	37, 60, 61
Khedive (P. 103)	29, 34, 90, 95, 99, 109	*Pierrot* (F. 143), 636	25, 39, 40, 62, 68, 105	Stoke Pogis, 3d, 2238	107
Koffee (F. 233)	111	*Pilot*, 3	59	*Sultan* (F. 58)	8, 25, 27, 28, 31, 34, 43, 52, 80
Landseer (P. 162)	85, 88, 89	Prince Hammond, 3672	91, 92, 99	*Sultane* (P. 7)	34
Lady Mary, 1148	40, 41, 42, 65, 70, 102	Prince of Wales	8, 25, 27, 28, 31, 34, 43, 49, 50, 52, 80	*Superb* (F. 353)	8, 25, 27, 28, 43, 49, 50, 80, 114
Lily, 1	98			Sweepstakes Duke, 1905	See *Duke* (P. 76)
Litchfield, 674	94, 98	Queen of the North (F. 1551)	19	*Tormentor* (F. 259)	29, 90, 95, 109
Lord Lisgar, 1066	59, 66, 107	*Rajah*, 340	54, 55	*Victor* (P. 148)	82, 84
Loyal (P. 70)	100	*Regina* (F. 32)	28, 52	Victor Hugo, 197	59, 66, 107
Lurlie Maid, 4159	57	*Rioter*, 2d, 469	13, 69	*Violet of Darlington*, 5573	11
Majoram, 3239	107	*Rival* (P. 143)	72	*Welcome* (P. 172)	20, 22, 24, 25, 33, 38, 43, 46, 48, 50, 73, 81, 91
Mercury, 432	68, 69, 95	*Romulus* (P. 181)	48, 50		
McClellan, 25	64, 66, 67, 94, 98	*Royalist* (P. 139)	28, 52	*Welcome* (F. 207)	92, 94
Musique (F. 1096)	24, 25, 48, 50	Sarpedon, 930	12, 13, 69	*White Rose* (F. 593)	10, 32, 103
Niobe, 99	41, 57, 66, 92	Satin, 10329	13	*Yankee* (P. 27)	10, 12, 32, 36, 63, 100, 103